Technical Meeting on Development, Qualification and Affordability of Complex Systems 2022

AA006312

Huntsville, Alabama, USA
14 – 15 September 2022

ISBN: 978-1-7138-6505-6

Printed from e-media with permission by:

Curran Associates, Inc.
57 Morehouse Lane
Red Hook, NY 12571

Some format issues inherent in the e-media version may also appear in this print version.
These proceedings were created from a scanned original document, and are the best quality available.

Copyright© (2022) by Vertical Flight Society
All rights reserved.

Printed with permission by Curran Associates, Inc. (2023)

For permission requests, please contact Vertical Flight Society
at the address below.

Vertical Flight Society
2701 Prosperity Ave, Suite 210
Fairfax, VA 22031
USA

Phone: (703) 684-6777
Fax: (703) 739-9279

www.vtol.org

Additional copies of this publication are available from:

Curran Associates, Inc.
57 Morehouse Lane
Red Hook, NY 12571 USA
Phone: 845-758-0400
Fax: 845-758-2633
Email: curran@proceedings.com
Web: www.proceedings.com

VFS Development, Qualification and Affordability of Complex Systems Technical Meeting Agenda
Sept 14–15, 2022
Radiance Technologies, Huntsville, AL

Wednesday, Sept 14, 2022		
0700-1200	**Attendee Badge Pick-up**	
0700-1200	**Presentation Material Drop-off for Authors**	
0800-0815	**Welcome by Redstone Chapter President** – Mr. Bill McCandless, US Army FLRAA Program Office	
	Meeting Objectives by General Chair – Mr. Eric Grigorian, GTRI	
0815-0900	***Opening Keynote: MG Robert Barrie, PEO AVN***	
	Session I – Safety & Certification **Session Chair: Dave Cripps, US Army CCDC AvMC, SRD**	
0900-0910	Safety and Certification Initial Comments: Dave Cripps	
0910-0930	**Session Invited Topic:** *IFR & IMC Qualification,* Mel Johnson, US Army CCDC AvMC	
0930-0955	*Certification Considerations for Flight Critical Complex Non-traditionally Developed Software,* Jason Rupert, Modern Technology Solutions	1
0955-1010	**Break**	
1010-1035	*Validation of Mission Performance Calculator (MPC) for Airworthiness Applications of Future Vertical Lift Platforms (Presentation Only),* John Sims, US Army DEVCOM AvMC	
1035-1100	*Safety and Product Robustness in the Air Vehicle Design Process,* Barton Hainsworth, Ide Ehigiato, Pete Barner, Sikorsky, a Lockheed Martin Co.	13
1100-1145	Questions and Roundtable Discussion	
1145-1230	**Lunch**	
	Session II - Modeling and Simulation **Session Chair: Alex Boydston, US Army CCDC AvMC**	
1230-1240	Modeling and Simulation Initial Comments: Alex Boydston	
1240-1300	**Session Invited Topic:** *SCADE/Transitioning to MBSE & Understanding ROI,* Justin Pearson, US Army CCDC AvMC	
1300-1325	*Modern Aircraft Flight Control Design and Bare Aircraft Sizing: an Integrated Model-Based Approach,* F.D. Amato, N. Cortigiani, D. Del Gobbo, L. Belluomini, D. King, Leonardo Helicopters	26
1325-1340	**Break**	
1340-1405	*Leveraging Digital Engineering to Reimagine Computational Analyses,* Dave O'Brien, US Army DEVCOM AvMC	39
1405-1430	*Analyzing Mechanical Systems Using 3DX Design Tables, Kinematic Simulations, and Dynamic Simulations,* Jack Faysash, Sikorsky, a Lockheed Martin Co.	45
1430-1515	Questions and Roundtable Discussion	
1515-1600	*Development, Qualification and Affordability of Complex Multicore Systems (Presentation Only),* John Warther, Green Hills Software	
1600-1730	**Social**	

VFS Development, Qualification and Affordability of Complex Systems Technical Meeting Agenda
Sept 14–15, 2022
Radiance Technologies, Huntsville, AL

Thursday, Sept 15, 2022		
0700-1200	**Attendee Badge Pick-up**	
0700-1200	**Presentation Material Drop-off for Authors**	
0800-0815	**Welcome by General Chair:** Mr. Eric Grigorian, GTRI	
0815-0900	***Opening Keynote:*** ***Mr. Dan Bailey, Deputy Director, Future Vertical Lift CFT***	
	Session III – Complex Design & Control **Session Chair: David O'Brien, US Army DEVCOM AvMC**	
0900-0910	Complex Design & Control Initial Comments: David O'Brien	
0910-0930	**Session Invited Topic:** *ERDC, HPC-enabled Design and Analyses,* Dr. Ian Dettwiller, U.S. Army Engineer Research & Development Center (ERDC)	
0930-0955	*Multidisciplinary Design of a Complex Tiltrotor Morphing Wing System,* D. Rogers, C. Punzi, A.A. Trezzini, P. Abdel Nour, Leonardo Helicopters	54
0955-1010	**Break**	
1010-1035	*Verification of Electric Aircraft during Steady State Operation and Fault Conditions (Presentation Only),* Mitchell Marks, HBK	
1035-1100	*Simulation and Integration Laboratory (SaIL) Testing of Integrated Cockpit and Fly By Wire,* Luca Belluomini and Sid Xiao, Leonardo Helicopters	63
1100-1145	Questions and Roundtable Discussion	
1145-1230	**Lunch**	
	Session IV – Lifecycle Digital Engineering **Session Chair: TBD**	
1230-1240	Digital Engineering Initial Comments: TBD	
1240-1300	*A Missing Piece of PEO Enterprise Architectures: The Assured Infrastructure Model,* M. Brown, D. Walsh, S. Nagel, H. Tiedeman, Collins Aerospace	80
1300-1325	*Component Specification Model,* Dr. Jon Fox, GTRI	
1325-1340	**Break**	
1340-1405	*Measuring Power Losses in Motors and Inverters (Presentation Only),* Mitchell Marks, HBK	
1430-1455	*Oil Debris Monitoring: Advanced Analytics Enabling Maintenance Efficiencies (Presentation Only),* Adam Oszmian, Gastops	
1455-1530	Questions and Roundtable Discussion	
1530-1600	**Closing** – Bill McCandless	

Certification Considerations for Flight Critical Complex Non-traditionally Developed Software

Jason K. Rupert
Senior Airworthiness Engineer
Modern Technology Solutions Inc.
Huntsville, AL 35806

ABSTRACT

When trying to reconcile the addition of non-traditionally developed software (e.g., Artificial Intelligence (AI), machine learning (ML), and other complex software) to flight and safety critical applications, the question of certification of the non-traditionally developed software is a difficult one to address and answer. First, the guidance for what might be expected for certification of such software is often unknown or in development, where the expected processes, objectives, and activities are uncertain. Second, there is a question of whether certification, even once those processes, objectives, and activities are identified, will be adequate or even possible given some of the potentially unique and non-deterministic aspects of the non-traditionally developed software. Third, instead of certification can mitigations [e.g., interlock, inhibit, interrupt (i.e., run-time assurance, or doer/checker), or limiter] only be applied, and therefore forego certification all together of their non-traditionally developed software. This paper explores the third question, while also considering the previous two: (1) what might be expected of non-traditionally developed software, and (2) adequacy of those expectations.

INTRODUCTION

This paper begins by defining traditionally developed software, non-traditionally developed software, and covering a few key differences between the two. After exploring that topic, a discussion of the civilian airworthiness assurance and safety authority expectations for traditionally developed software is provided. Then the paper explores four types of mitigations, and finally the recommended assurance expectations for non-traditionally developed software are covered. Each of these sections provides unique supporting evidence indicating that full assurance processes, objectives, and activities are necessary for the safe and airworthy addition of non-traditionally developed software to flight critical applications. The discussed assurance processes, objectives, or activities aggregate together to jointly ensure the possibility of non-traditionally developed software contributing to the introduction of risk is lowered or removed. Failing to complete any one of those items introduces and increases the possibility of non-traditionally developed software contributing to a resulting hazard/mishap.

Risk is calculated as a multiple of probability of occurrence of the hazard/mishap and severity of outcome of the hazard/mishap. The complexity and probabilistic nature of the non-traditional development process adds to the probability of failure, and thus increase the risk associated with the use of non-traditionally developed software. This paper explores the use of assurances and mitigations to reduce that risk, where mitigations are independent and developed to the appropriate assurance level. Through the application of the assurances and mitigations the risks will be reduced thru reduction of the probability of occurrence of the event.

Certification is the process of presenting an airworthiness case and receiving appropriate releases by the proper safety and airworthy authorities that provide flight and materiel release for flight and/or safety critical non-traditionally developed software component.

This paper concludes by not discounting the use of mitigations, but instead emphasizing the necessity of the assurance process to establish the boundaries of the systems, conduct necessary assurances processes, objectives, and activities, and then consider areas where additional risk remains that should be mitigated. This approach was true for traditionally developed software and remains true for non-traditionally developed software. That is, mitigations, when developed to the proper assurance level and are independent of the function, to reduce the probability (control category) of the failure, just as assurance techniques are also applied to reduce the presence of faults and failures.

Presented at the Vertical Flight Society's 2022 Development, Qualification and Affordability of Complex Systems Technical Meeting, Huntsville, AL, USA, September 14-15, 2022, Copyright © 2022 by the Vertical Flight Society. All rights reserved.

TRADITIONALLY VS. COMPLEX NON-TRADITIONALLY DEVELOPED SOFTWARE

Traditionally Developed Software

As shown in Figure 1, traditionally developed software is source code (implementation) that can be explicitly and fully developed from textual or graphical software low-level requirements, architecture, and design data items. Those data items contain all information necessary to develop the algorithmic behavior of the source code including parametric value settings to be developed. The low-level requirements are derived from software high-level requirements (HLRs), which are derived from systems level requirements. Important are the three levels of requirements decomposition associated with traditionally developed software development, and the bi-directional traceability that exists between the three levels.

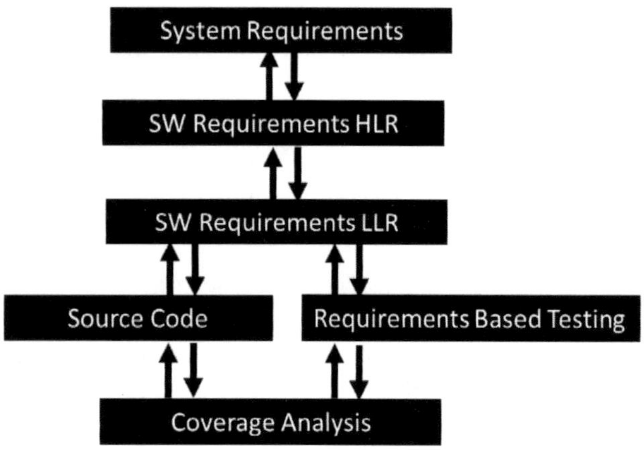

Figure 1. Bi-directional traceability for traditionally developed software development process.

Moreover, the same low-level requirements (LLRs) used for the development of source code are also used to develop the verification plans and procedures. The verification plans and procedures are used to conduct requirements-based testing and source code coverage analysis. These are fundamental characteristics of traditional source code that allow full bi-directional traceability between software requirements, source code, and verification cases. Bi-directional traceability illuminates areas where source code exists that should not, where test cases are missing, and areas where requirements (or architecture elements) were omitted during the requirements decomposition process.

For the purposes of this paper, the term verification is defined as follows: "The evaluation of outputs of a process to ensure correctness and consistency with respect to the inputs and standards provided to that process." (Ref. 1).

Complex Non-traditionally Developed Software

Complex non-traditionally developed software, especially data-driven machine learning software, typically does not use traditionally developed software development methodologies, e.g., complex non-traditionally developed software does not develop implementation source code and parameter values directly from low-level requirements (LLRs). Instead, as shown in Figure 2, data-driven methodologies are used to train machine learning algorithms to perform classification or regression determinations, where training consist of using automatic techniques to assign hyperparameter values. That is an oversimplification of the approach, but that description serves the purpose to indicate that no longer are LLRs, which capture algorithm detail, being used to develop the source code, and set parametric values. Instead, machine learning requirements consist of model and data requirements, are being used to determine the best and establish the machine learning (ML) architecture.

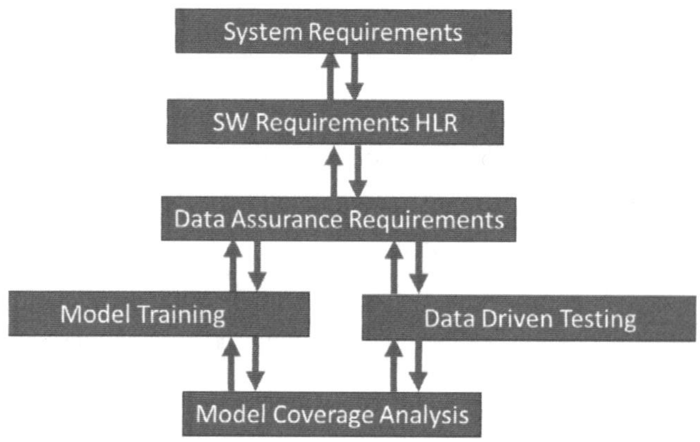

Figure 2. Possible bi-directional traceability for non-traditionally developed software development process.

The ML data requirements are then used to drive the ML data collection process, which could involve using existing curated tagged data sets, or pursuing the process of establishing the collection and tagging of the data from natural environment, synthetic environment, or a combination of sources. In addition, to the collection of the data, tags and enumerations must be added to the data set to indicate the attributes on which the ML model is to conduct the learning process. The tags and data source must be synchronized so that the tags are synched with the data set. In the case of textual based data source [e.g., comma separated values (CSV) files] this processing and curation is manageable with commonly available tools. However, identifying and attributes in streaming video sources from multiple angles and sensors, and synchronization process those multiple streams requires specialized tools. The previous describes the machine learning data collection and processing process, which is already a deviation from the traditionally developed software develop process. Going forward the machine learning development processes only continues to deviate from the traditionally developed software development process. Further data extraction, transforming and loading (ETL) and other data science related manipulation may be necessary to align the data with a usable format.

The training process, where an independent training data set is established from the collected and processed data set, is one more example where the non-traditionally developed software process deviates. During the ML development process the training data set is used to determine the ML model parameter and hyperparameter values.

The whole data set is collected to satisfy the ability of the machine learning model to fulfill regression or classification decision for a specific operational design domain (ODD), where ODD is defined as the following:

"The set of environments and situations the item is intended to operate within. This includes not only direct environmental conditions and geographic restrictions, but also a characterization of the set of objects, events, and other conditions that will occur within that environment." (Ref. 2)

The operational design domain (ODD) may be simple in the case of predicting failure of a single sensor based on trend data or incredibly complex in the case of identifying landing zone obstacles (like wires) in degraded flight environments. Example additional considerations for defining an ODD are provided in "How Many Operational Design Domains, Objects, and Events?" (Ref. 3) In both cases enough (e.g., volume) and appropriate (e.g., adequate) data will need to be collected to construct three independent sets of data: training, verification, and test, where those three data sets are independent and appropriate to ensure the model does not inadvertently acquire information it should not have, which is called leakage.

The data sets will need to be large enough to guarantee that overfitting and underfitting are not occurring within the model, and that the full ODD is covered to the extent that robustness testing can occur to ensure the machine learning model will respond successfully to anticipated failure cases. Unanticipated failure cases should also be handled through systems mitigation processes. Overfitting is where the machine learning model is large/complex enough (e.g., millions of neurons) that it can memorize the training data set. Overfitting is normally uncovered when the verification and test data sets are used, and the model performance degrades from 100% accuracy on training data to unacceptable levels for the verification and test data sets. Underfitting occurs when the data set is too small, which leads to poor training performance where not enough information is available to properly set hyperparameter values to be able to classify objects or predict outcomes. In the case of underfitting, larger datasets are needed, and most likely a more complex machine learning model architecture should be developed. The use of the training data set to produce the desired functional behavior is used by the Navy and Army safety community as a key discriminator between traditional and non-traditional developed software:

> "The hypothesis is that a function is an AI [Artificial Intelligence] Type Function candidate if one or both criteria are met:
>
> Criteria 1 - Data Approximations. The function requires the use of data approximations to build/train its algorithm. …
>
> Criteria 2 - Data Samples. The function requires the use of data samples to build/design its algorithm. ..." (Ref. 4)

Not all AI techniques use data-driven training techniques. However, much of the aviation and commercial industry sees that AI techniques that use data-driven training techniques are those most likely to be used in most commercially viable applications. Convolutional neural network (CNN) and deep neural network (DNN) are prominent types of data driven AI techniques, where machine learning (ML) is used synonymously with AI data driven learning techniques. Techniques outside of data driven machine learning are beyond the scope of this paper as they can be covered by traditional software development techniques or are not mainstream enough to be considered commercially viable. The one exception is reinforcement learning. For reinforcement learning an independent assessment will be conducted outside the scope of this paper to examine the applicability of the processes, objectives, and activities proposed here to reinforcement learning and propose additions where the assurance approach discussed below falls short.

The machine learning model may ultimately be operating in an environment where there may be many anticipated unknowns-unknowns, i.e. unexpected features that may occur but are not able to be replicated in the collected data set (either through simulated/synthetic or naturally collected means). In those cases, since they are not able to be replicated, the behavior of the machine learning model response is unknown. For those unknown cases robustness in the system and subsystem design is necessary.

Continuous learning (Ref. 5) (adaptation) is not recommended at this time because robustness of the assurance cases of the system and/or subsystem incorporating the machine learning model should be predictably and verifiably safe and airworthy. Continuous learning (adaptation) is a non-traditionally developed software development approach where the machine learning algorithm is allowed to continuously sample, expand the training data set, and then use the expanded training data set while online (deployed) to update the machine learning model's own hyperparameters. In most cases the data set collected and then used for continuous learning (adaption) has not been supervised or reviewed nor have consequences of the adaption been fully considered. Given these uncertainties, continuous online (adaptive) learning is not recommended due to the extensive online regression, verification, and behavioral uncertainties that results from adaption occurring while the system is fielded. Instead, inferences models and their hyperparameters are recommended to be locked (Ref. 5) (frozen) when deployed to not allow online continuous learning (adaption).

Assurances are necessary that the system and/or subsystem with the machine learning non-traditionally developed software will fail safe or respond appropriately when exposed to stimulus from outside the defined operational design domain (ODD). All steps should be taken to ensure the machine learning data requirements, operational design domain (ODD), machine learning data collection and data processing activities are as complete as possible and all due diligence is pursued to account for anticipated known unknowns, i.e., unlikely but possible deviations anticipated in the operational design domain environment, e.g., flying through a large group of bats or cicadas.

During wartime the anticipated ODD is may change beyond those cases anticipated, so ensuring the system and subsystem incorporating responds safety is essential. Should the machine learning data requirements, operational design domain, machine learning data collection or data processing activities neglect to include off-nominal cases through omission (oversight) the outcome is the same as a gap in in traditional systems and/or software requirements development process. This gap will be uncovered by executing traceability, i.e. machine learning data collection and data processing activities trace to operational design domain machine learning data requirements, and operational design domain machine learning data requirements trace to systems level operational design domain requirements. This traceability highlights the importance of the systems engineering process adequately developing the operational design domain (ODD) requirements, because without those being complete and of high quality the collection of the data for the machine learning data processing activity will be inadequate. Figure 2 may show, in comparison with the traditionally developed software development process shown in Figure 1, the bi-directional traceability and verification activities that may be necessary to produce assurance for the development of non-traditionally developed flight-critical software. Figure 2 is not based on any specific source material, but instead was put together to help illustrate the types of traceability that will be necessary for non-traditionally developed software. Following sections explore these topics further with references to specific standards exploring the development of assurance processes, objectives, and activities for non-traditional developed software.

CIVILIAN AVIATION APPROACH FOR TRADITIONALLY DEVELOPED SOFTWARE

The civilian software airworthiness assurance approach determined the best way to gauge the assurance expectations for software was to establish software item design assurance expectations. Those expectations (levels) are established using hazard assessment activities, e.g., functional hazard assessment (FHA), fault tree assessment (FTA), failure modes effects and criticality analysis (FMECA). Higher development assurance levels indicate that greater processes, objectives, and activities should be accomplished. The process for accomplishing hazard assessments, which involves assessing the airplane system and subsystem functionals and functional requirements, are detailed within the following:

- SAE International Aerospace Recommended Practice (ARP) 4754 Guidelines for Development of Civil Aircraft and Systems (Ref 6.), and
- SAE APR 4761 Guidelines and Methods for Conducting the Safety Assessment Process on Civil Airborne Systems and Equipment (Ref. 7).

Once the system and subsystem hazard assessments are complete, the design assurance level (DAL) flows to the software, hardware (electronics), and integrated modular avionics (IMA) lifecycle process, respectively.

In general, the number of processes, objectives, and activities necessary at the systems, software and complex electronics level are proportional the DAL assessed for each, so for higher DAL a greater number of expected processes, objectives, and activities should be accomplished. The DAL is established at the systems and subsystems level based on the hazard assessment and functional decomposition. As shown in Table 1, and discussed extensively in SAE ARP 4754 and SAE 4761, DAL are assigned based on top-level failure condition classification severity of the function, where DAL A is the highest and associated with a Catastrophic failure condition of the function.

Table 1. Determining DAL Assignment.

DAL	Failure Condition Severity	Severity Aircraft Effect	Effect on Crew
A	Catastrophic	Complete loss of aircraft	Crew unable to accomplish required tasks, or Required crew strength or skill in excess of crew capability, or Crew incapacitation, or Crew fatalities
B	Hazardous	Large reduction in aircraft functional capability or safety margin	Excessive crew workload increase or Crew unable to fully accomplish required tasks, or Crew physical distress
C	Major	Significantly reduced aircraft functional capability or safety margin	Significant crew workload increase or Conditions impairing crew efficiency, or Crew physical discomfort
D	Minor	Slightly reduced aircraft functional capability or safety margin	Slight crew workload increase
E	No Safety Effect	No effect on aircraft functional capability or safety margin	No effect on crew workload or physiology

Note: For systems containing non-traditionally developed software, traditional safety assessment approaches, like FHA, FTA, and FMECA, may not be adequate, and more sophisticated safety assessment tools may be necessary for assessing complex systems automated behaviors that may be interacting with humans or other non-traditionally developed software. In those cases, unlike the traditional safety tools, modern safety assessment tools like system-theoretic model of accidents (STAMP) and System-Theoretic Process Analysis (STPA) by Dr. Nancy Leveson may be more appropriate since they are able to account for complex, probabilistic, and non-deterministic interactions (Ref. 8).

Accomplishing the processes, objectives, and activities fully (completely and adequate quality) ensures that the appropriate level of design, development and verification rigor has been applied to reduce/eliminate the risks associated with the traditionally developed software. For high development assurance level (DAL) software items, the adequacy is established by putting in place the appropriate plans and standards, developing high-level requirements (HLRs) and low-level requirements (LLRs), building the software architecture and implementation (code), and conducting appropriate verification all those elements. These processes, objectives, and activities are fully detailed in RTCA DO-178C. There is a prorated levelling relationship for the number of DAL objectives to the DAL, so that as the DAL decreases, i.e., from DAL A to DAL D, so do the expected number of objectives to be accomplished, e.g., for DAL A there are 71 objectives and for DAL D there are 26.

Impacts Of Non-Traditionally Developed Software

Certifying non-traditionally developed software will be one of the greatest struggles the assurance discipline has encountered since the early 1980s, i.e., when traditionally developed software was initially added to the flight and safety critical assurance discipline. The first version of DO-178 Software Considerations in Airborne Systems and Equipment Certification was published in November 1981 (Ref. 9). While most of the content and structure would not be recognizable to those familiar with the current version of DO-178C , the goal of the standard was to "develop and document software practices that would support certification of software-based equipment and systems". DO-178 (1981) indicates:

> "The aviation community has many years of experience in the design, production, installation and use of equipment and systems based on analogue technology, including systems performing flight-critical functions...In the future, digital computer-based equipment and systems will be the rule rather than the exception. Unlike previous digital systems, they will be deployed in new system architectures and will communicate digitally. They will also be applied to flight-critical functions, and for systems whose essential safety is not provided by analog monitors or authority-limiting devices, the quality of their software may relate directly to flight safety." (Ref. 9)

"in the future" has been reached and software based digital systems are performing the highest criticality functions on manned and unmanned airborne systems, e.g., environmental control, engine control, flight control, etc. Those flight-critical software-based systems are monitored by other dissimilar software-based systems and will be relied upon to operate in failover situations (with no mention of analog systems).

The 1981 version of DO-178 continued to mature and adapt to the evolving and maturing software development techniques. For example, in the last decade the aviation assurance community embraced modern software development technologies and techniques by updating to DO-178C and adding its four supporting supplements, i.e., RTCA DO-330 (Ref. 10), RTCA DO-331 (Ref. 11), RTCA DO-332 (Ref. 12), and RTCA DO-333 (Ref.13). The addition of DO-331 and DO-332 provide guidance for the certifiable use of model-based development (MBD) and object-oriented (OO) approaches and technologies on flight and safety critical applications, respectively. The aviation assurance community has successfully used an evolutionary approach to mature the DO-178 content, so this established precedent for maturing standards can be leveraged for updating foundational standards to current technologies, e.g., inclusion of non-traditionally developed software in the aviation standards framework.

By appropriately updating and formalizing the expected software lifecycle processes, objectives, and activities for non-traditionally developed software, the aviation assurance community can place confidence that non-traditionally developed software can safely perform performing flight and safety critical functions. For software assurance to appropriately address non-traditionally developed software, enhancements are necessary to the current foundational assurance documents (Ref. 14 and Ref. 15). As shown in Figure 3, every facet of software assurance from configuration management to verification to sustainment and maintenance is impacted by the addition of non-traditionally developed software, e.g., AI and ML. As shown in Figure 3, new areas (shown in dark blue/dark green polygons) must be added to the existing traditional assurance areas (shown in red/green polygons) to account for non-traditionally developed software. In Figure 3, AI is used to represent one type of non-traditionally developed software, but other types can also be considered.

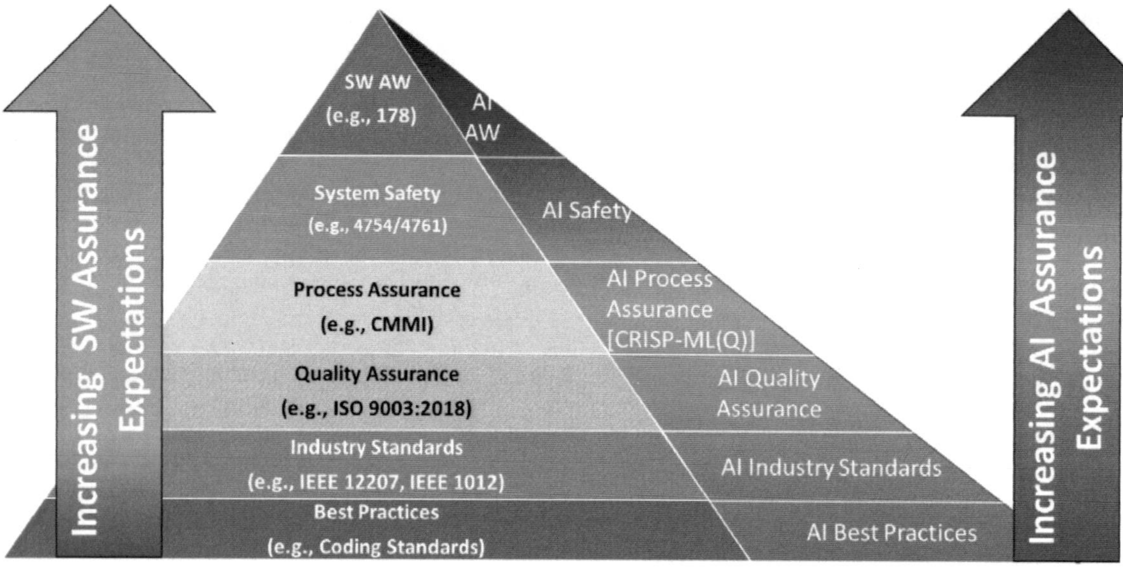

Figure 3. Non-traditionally developed software extending software assurance processes.

MITIGATIONS

For non-traditionally developed software, mitigations may struggle in their ability to cover the full and complex ODD where a non-traditionally developed software element might be expected to operate. If mitigations are used, i.e., hazard is not able to be designed out of the system, extensive verification efforts must be planned ensure mitigations safely mitigate hazards in complex ODDs. Moreover, the safety hazard assessment and level of rigor processes, objectives, and activities should further substantiate the mitigations adequately ensure a break in the mishap chain, and the mitigation is independent and developed to the appropriate assurance level. In most cases, following those processes should drive the hazard and risk out of the system, or reduce it to an acceptable level given the mission and function. Driving the hazard and risk out of the system by design choices is the #1 recommended way to eliminate risk by safety assurance organizations:

- FAA System Safety Handbook (See Table 3-7 "Safety Order of Precedence") (Ref. 16),
- MIL-STD-882E (Section 4.3.4) (Ref. 17), and
- US ARMY Regulation (AR) 385-10 (Section 9-2.j) (Ref. 18).

Interrupt

An interrupt mitigation is designed to break the mishap sequence, e.g., function capable of stopping ongoing weapon firing when safety constraints are reached. While there are many types of interrupts (hardware, software, and electrical), two types are germane to non-traditionally developed software: run-time assurance, and doer/checker. Unfortunately no specific definition of an interrupt is provided in MIL-STD-882E, Joint Software System Safety Engineering Handbook (JSSSEH) Implementation Guide (Ref. 19), or Army Military Airworthiness Certification Criteria (AMACC) (Ref. 20). Interrupt involves the use of a monitor as well as a follow up action that executes once the monitor detects the fault or hazard is likely or eminent, e.g., the interrupt could stop the designator laser from firing if an aircraft warning or caution status is detected. A sophisticated application of an interrupt mishap can be thought of one that involves the ability to stop the progression toward the hazard and regress the function to a safe state, mode, or condition.

Interrupt Example

Run-time assurance (RTA) is most definitively addressed in ASTM 3269-17 "Standard Practice for Methods to Safely Bound Flight Behavior of Unmanned Aircraft Systems Containing Complex Functions" (Ref. 21), where RTA is defined as:

> "implements real-time monitoring, prediction, and fail-safe recovery mechanisms that bounds the flight behavior of a non-pedigreed complex function to ensure the safety of a UAS." (Ref. 21)

That is, RTA provides functionality like the limit and interrupt functionality discussed in the AMACC for the Weapon Inhibits, Limits and Interrupts (WILI) while also providing the interlock/inhibit like functionality, i.e., safety control mechanism that stops the behavior from occurring before it occurs.

Programs attempting to use RTA functionality, e.g., NASA EVAA/Traveler , have uncovered other weaknesses with the RTA infrastructure, e.g., the baseline ASTM 3269-17 framework did not account for the case where multiple hazards are monitored and multiple simultaneous recoveries planned. In that case, NASA created the multi-monitor RTA, i.e., MM-RTA. For a MM-RTA, a consequence arbitrator chooses which recovery function to implement based on which outcome has the highest consequence score. However, the consequence arbitrator, which NASA calls Moral Compass (although no morality is involved in the maximum consequence selection), is inadequate alone, and led to the addition of the crosstalk channel from the ground collision calculation to the airborne collision calculation. These necessary enhancements (as well as several others) to baseline ASTM 3269-17 RTA framework to cope with just these two easily conceptualized monitors and recovery functions illustrates the challenges ahead for apply RTA approaches to bound non-traditionally developed software. Adding more monitors and recovery functions that are expected to function sequentially or simultaneously will lead to increasingly intertwined architecture.

Figure 4 presents an example MM-RTA architecture that would allow multiple monitors to simultaneously monitor hazards. The Moral Compass, shown in Figure 5, allows for the appropriate recovery function to be selected to address the highest consequence hazard identified by the safety monitors. The Crosstalk channel, shown in Figure 4, allows some coordination between recovery functions to allow simultaneous hazards to be considered, e.g., airborne collision avoidance to also consider ground collision avoidance, and avoided. Modern Technology Solutions, Inc. is executing such an architecture in their Supervised Autonomous Function Executive (SAFE®) MM-RTA architecture.

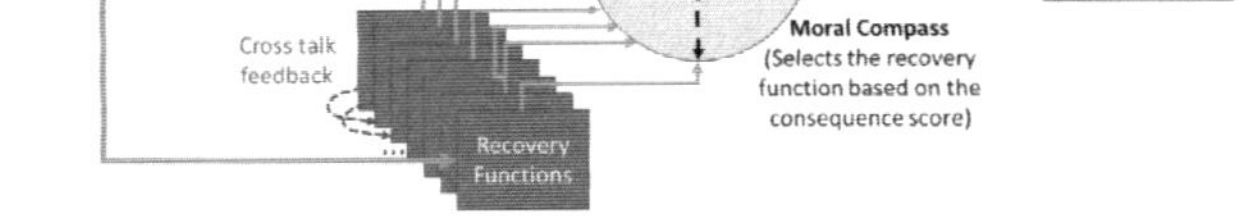

Figure 4. Example MM-RTA Architecture.

As complexity of the MM-RTA increases, more crosstalk between monitors and recovery functions may be necessary, and potentially nesting of recovery functions.

Noteworthy for the approach discussed in NASA/CR–2020-220586 (Ref. 22), is that certification of the complex software is not to avoided but instead RTA is used to enable the test, verification, and certification of the complex software, i.e., the certification processes, objectives, activities, and data items for the complex software to be pursued in parallel with the flight test program.

> "During flight, while the high-automation algorithms are operating, the RTA system will monitor the aircraft state for any impending violation of safety requirements. When necessary, it will switch to the low-automation software to prevent such violations. Because the pilots are highly trained, this switch in control could be reasonable, provided that the pilots are familiar with this switching behavior and such a switch is properly announced. The RTA system will be assigned a high DAL, commensurate with that of the low-automation control algorithms... [for example] if operated by a skilled pilot, trained to safely fly the aircraft in ACAH (Attitude Command/Attitude Hold) mode, the TRC (Translation-Rate Control) mode could be paired with a separate ACAH mode and RTA components that switch control modes as appropriate. This could significantly reduce the burden of V&V for the TRC algorithm at the cost of V&V for the RTA components." (Ref. 22)

Moreover, within NASA/CR–2020-220586, a qualified experienced test pilot is in the loop with the complex software that is also monitored by the RTA recovery function. The test pilot's role is to use the complex software in development. The monitor can gauge, along with the test pilot, when the complex function is not performing adequately and determine failover to the traditional RTA recovery function is necessary. When that failover occurs, the traditionally developed RTA recovery function is under the control of the experience pilot.

AIRWORTHINESS AND SAFETY ASSESSMENT

As alluded to above, answering the question of whether mitigations alone will be adequate is impossible without first performing an appropriate safety assessment, where the safety assessment will consist of at least an FHA, FTA, and FMECA. The FHA should be performed at the aircraft system and subsystem level and be used to guide the development of the FTA. Cutsets from the FTA will then be used to guide the FMECA. Each of these will be used to determine the DAL of each function and identify where derived safety requirements and functions are necessary at the systems and sub-systems level. Once identified, those assessments transitioned along with the system/subsystem requirements allocated to the software item. Within the software development lifecycle, the software item will be developed to the appropriate DAL. Developing to the appropriate DAL, i.e., accomplishing the objectives indicated in Table 2 and producing the data items shown in

Table 3, ensures the software item itself is not contributing its own hazards. Through the combination of the assurance development and the addition of system/subsystem mitigations (building blocks), with the appropriate system/subsystem verification (not addressed in this paper), assurance is provided.

Airworthiness and Safety - DAL and LOR Development

The military safety and software safety processes (i.e., MIL-STD-882E, AMCOM Regulation 385-17 (Ref. 23), and JSSSEH Implementation Guide) follows a similar course by requiring the execution of the system decomposition, safety assessment, and assignment of the LOR (like DAL), and then the software item developed to the appropriate LOR. MIL-STD-882E is more explicit on the penalty of not completing the development the software item to the appropriate LOR by providing Table 7, which specifies the risk due to "unspecified or incomplete" LOR objectives and activities, i.e., increased risk due to the software component.

Additional risk occurs because by not following the processes, objective, and activities to completeness no longer can the airworthiness authorities have confidence that the software was developed to comply with the parent system requirements. DO-178C states that its' purpose is the following: "provides the aviation community with guidance for determining in a consistent manner and with acceptable level of confidence that the software aspects of airborne system and equipment comply with airworthiness requirements." By not following the processes, objectives, and activities within safety and airworthiness guides (i.e., AMCOM Regulation 385-17 and RTCA DO-178C, respectively), no longer will that be true. Confidence is lost that the software was adequately developed to comply with airworthiness or any requirements. This is the crux of the issue when contemplating the question of solely relying on mitigations [e.g., interlock, inhibit, interrupt (i.e., run-time assurance, or doer/checker), or limiter] alone to justify the use of non-certified non-traditionally developed software in flight and safety critical applications. Without including design assurances, certifying authorities would have no confidence in the non-traditionally developed software or its form, fit, and function for its desired flight and safety critical application. Moreover, appropriate assurance cases are necessary for mitigations as well. For complex mitigation a great amount of effort is required to building up a safety case for the mitigations [e.g., interlock, inhibit, interrupt (i.e., run-time assurance, or doer/checker), or limiter]. To build a safety case where a mitigation alone could mitigate a flight and safety critical failed functions from the non-certified non-traditionally developed software without design assurances (e.g., processes, objective, activities, or data items) would be very difficult (if possible). Hopefully, the consideration is becoming unnecessary.

Non-traditionally developed software processes, objectives, and activities

One final justification that may be cited for avoiding attempting to certify non-traditionally developed software is that there are no non-traditionally developed software processes, objectives, and activities. Prior to 2020, an organization may have been able to argue this position. However, since 2020 there has been an explosion of legitimate guidelines indicating processes, objectives, and activities that non-traditionally developed software development programs should follow:

- Flight and Safety Critical Applicable
  - Safety Assurance Objectives for Autonomous Systems. (Ref. 24)
  - European Union Aviation Safety Agency (EASA) Concept Paper: First usable guidance for Level 1 machine learning applications. (Ref. 25)
  - Process Standard for Development and Certification/Approval of Aeronautical Safety-Related Products Implementing AI. (Ref. 26)
  - Standard for Safety for the Evaluation of Autonomous Vehicles and Other Products. (Ref. 2)
- Commercial Industry Focused Processes
  - Cognilytica Cognitive Project Management for Artificial Intelligence (CPMAI). (Ref. 27)
  - CRoss Industry Standard Process for Data Mining (CRISP-DM). (Ref. 28)
  - Microsoft Team Data Science Process (TDSP). (Ref. 29)

Other key publications indicate the errors to be expected should programs not follow fundamental best practices of processes, objectives, and activities:

- Traceability of Deep Neural Networks (Ref. 32).
- Software Engineering for Machine Learning: A Case Study (Ref. 32).
- Concrete Problems in AI Safety (Ref, 33).

These guidelines span all industry sectors and continue to be refined by active workgroups (e.g., SAE G-34, where participation is free) attempting to ensure non-traditionally developed software can be applied to flight and safety critical applications.

Airworthiness and Safety - DAL and LOR Development

As shown in Figure 5, the combination of the three processes listed above, i.e., SCSC-153 (Ref. 24), EASA Level 1 (Ref. 25), and AS6983 (Ref. 26), along with guidance from the Food and Drug Administration (FDA) (Ref. 34 and Ref. 35), i.e., Algorithm Change Protocol (ACP), Good Machine Learning Practice (GMLP), Predetermined Change Control Plan (PCCP), Software as a Medical Device (SaMD), and SaMD Pre-Specifications (SPS), can be used to produce an appliable process for certifying ML non-traditional flight and safety critical software applications. The FDA guidance

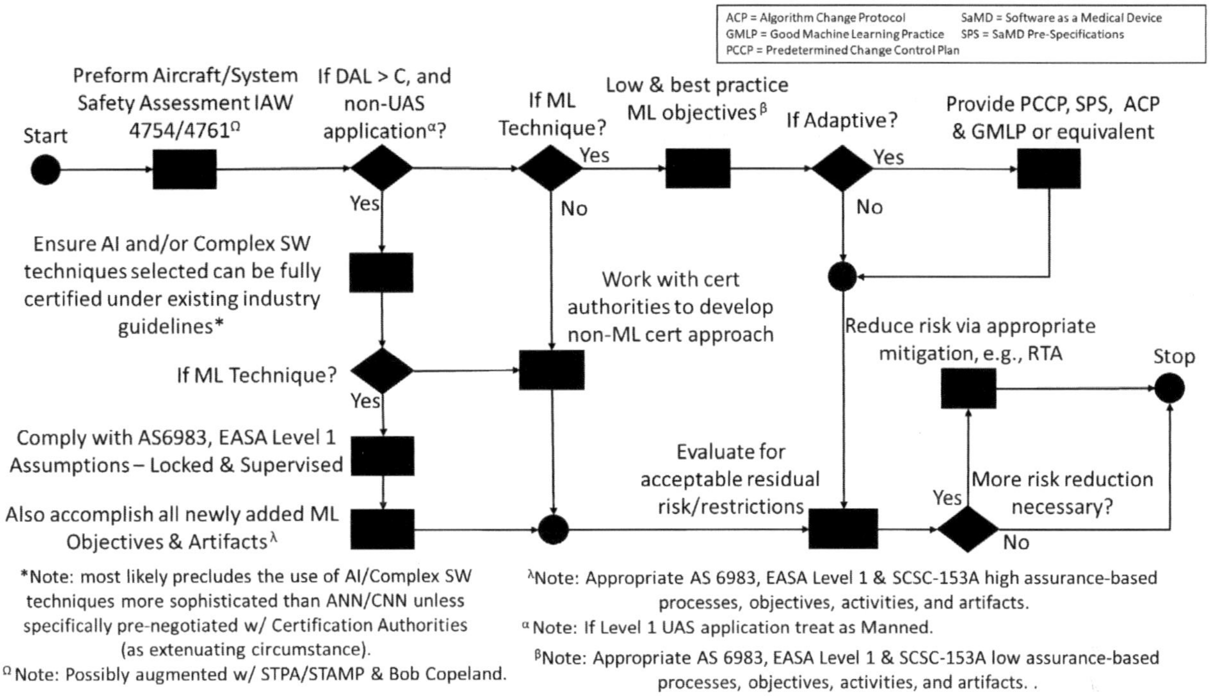

Figure 5. Possible Harmonized ML Certification Flowchart.

is unique because it attempts to allow the use of adaptive learning, which may be applicable for low-risk Army Regulation (AR) 70-62 (Ref. 36) "Airworthiness of Aircraft Systems" Airworthiness Qualification Level (AQL) 3 airworthiness unmanned aircraft system (UAS). FDA medical device and other similar terminology and concepts would need to be replaced with equivalent aviation guidance. In Figure 5, complex software (SW) is synonymous with non-traditionally developed software, e.g., machine learning (ML).

For Figure 5, highlights of the flow chart are the following:

- Begin following the traditional SAE ARP 4754/4761 process, but augmented with STPA/STAMP and other appropriate hazard assessment tools and techniques that accounts for the probabilistic nature of non-traditionally developed software,
- For manned flight and safety critical applications, ensure non-traditionally developed software development (e.g., AI and/or Complex SW) techniques selected can be fully certified under existing industry guidelines,
  - Ensure the ML development lifecycle follows a data-driven learning technique, e.g., Bayesian, Decision Trees, Artificial Neural Networks (ANN), Convolutional Neural Networks (CNN), and Deep Neural Networks (DNN).
  - Ensure the ML development lifecycle complies with locked/frozen learning, i.e., inference model is not adapting hyperparameter values while fielded, and the dataset was developed manually collected, produced, and tagged and then used to perform ML model training,
  - During the ML development lifecycle, the process, objectives, and activities should conform to applicable AS 6983 (processes, objectives, and activities), EASA Level 1 (objectives), and SCSC-153A (objectives), especially the processes that lead to the production of the complete and adequate ML Model Description,
  - Prior to beginning the creation of the ML item/implementation lifecycle process, which is also called the ML inference model lifecycle process, evaluate the performance of the ML model to identify if any additional risk and safety mitigations are necessary,
- As necessary reduce the risk appropriately through the development and application of appropriate mitigations [e.g., interlock, inhibit, interrupt (i.e., run-time assurance, or doer/checker), or limiter].

At this time, the ML inference model will be integrated with the system/subsystem, and appropriate subsystem/system integration verification activities conducted. The verification activities may uncover that additional ML functional enhancements are necessary, which would cause the ML process to be iterated appropriately. The above provides highlights of the hybrid process allowing for the certification of non-traditionally developed software.

Subsequent paper will provide greater detail and decomposition of these individual steps of the flowchart.

CONCLUSIONS

To allow for the certification of non-traditionally developed software the previous section introduced the highlights of what should be included in a hybrid certification framework. That framework extends and augments the traditional assurance framework to address the unique aspects of non-traditionally developed software. By extending assurances are provided that non-traditionally developed software containing flight critical software can responsibly include flight and safety critical functions and be added to flight and safety critical applications. No one additional output data item, process, objective, or activity will provide the full assurances, but instead a hybrid framework, including novel methodologies (e.g., STPA, simulation, mitigations), is necessary to accomplish the goal. That is, for non-traditionally developed (e.g., AI, ML, and complex software) flight and safety critical software applications the addition of mitigations [e.g., interlock, inhibit, interrupt (i.e., run-time assurance, or doer/checker), or limiter] alone will not be able to provide full assurances to certifying authorities and assurance stakeholder. When combined with an assurance framework, mitigations offer another technique to reduce the risk to an acceptable level.

To support this argument the following justifications were provided:

- Thorough safety assessment (accounting for probabilistic behaviors) is necessary to determine the level of rigor and design assurance level for each system, sub-system, hardware, and software item,
- Software items (traditional or non-traditional) should be developed to their appropriate level of rigor and design assurance, which establish the processes, objectives, and activities,
- For non-traditionally developed software, guidance for processes, objectives, and activities exists and can be leveraged from legitimate industry safety and airworthiness professional societies to indicate the types of processes, objectives, and activities that should be executed for non-traditionally developed software,
- Providing no certifying substantiation artifacts/data items (for traditional or non-traditionally developed software) causes a lack of confidence that the software item is executing the desired requirements, is fit for purposes, and is safe,
- Independent and appropriately developed (to the appropriate design assurance and level of rigor) mitigations types (especially when combined) can be valuable to reduce the risk of specific hazards,
- Only through the airworthiness and safety level of rigor and design assurance processes, objectives, and activities, which includes evaluations of mitigations, can adequate confidence be established,
- The assurance techniques (processes, objectives, and activities) will be augmented with novel approaches (e.g., Monte Carlo) to account for the probabilistic behaviors of non-traditional behavior, especially at the boundaries of its operational design domain.

Through further collaboration with stakeholders and cross functional teams, additional refinement of these concepts can occur. Only through additional collaborative interactions can weakness and gaps in these considerations be uncovered. No stakeholders want to see mishaps caused by the addition of non-traditionally developed software. While there is a rush for the inclusion of the technology on flight and safety applications all are working together to define the necessary amount of assurance. Through such collaboration a refined collection of output data items, processes, objectives, and activities can be established – and more importantly the rational and criteria for each understood.

No stakeholders are arguing mitigation alone as a solution but are instead working toward the appropriate balanced approach to assurance. That is, the assurance and aviation industry stakeholders, like the author's own experience, is that the mitigation alone approach proves to have limited applicability. Using mitigations alone breaks down in other flight and safety critical regions not fully anticipated. Moreover, using mitigations alone does not address underlying flight and safety critical lack of maturity (assurance), and just leads to a long, painful, and expensive series of fly-fix-fly-fix-crash-fix-fly "do loops". Going forward with a hybrid framework to address non-traditionally developed software will ensure assurance as well as reduced risk, where future publications will address the specifics of that hybrid framework.

Author contact: Jason K. Rupert jason.rupert@mtsi-va.com

DISCLAIMER

The views and opinions expressed in the attached are those of the author and do not necessarily reflect the views or positions of any entities they represent, i.e., does not provide the official position of U.S. Army Combat Capabilities Development Command (CCDC) Systems Readiness Directorate (SRD) with respect to establishing the Airworthiness Assurance argument for Artificial Intelligence and Machine Learning. The contents of this document are 100% based on a document that was successfully reviewed for Public Release (ID 045-22).

REFERENCES

1. *Software Considerations in Airborne Systems and Equipment Certification*, RTCA DO-178C, Prepared by: SC-205, RTCA, Inc., Washington, D.C., December 13, 2011.

2. *Standard for Safety for the Evaluation of Autonomous Vehicles and Other Products*, Standard 4600, Edition 1,

Edition Date: April 01, 2020.

3. Koopman, P., and Fratrik, F., "How Many Operational Design Domains, Objects, and Events?", Safe AI 2019: AAAI Workshop on Artificial Intelligence Safety, Jan 27, 2019.

4. Nagy, B. "Level of Rigor for Artificial Intelligence Development", Mission Systems Acquisition and Integration (MSA&I) Division, Software and Mission Systems Integration Department, Systems Engineering Department, April 2022, DTIC #: AD1173626.

5. ANSI/CTA Standard, Definitions and Characteristics of Artificial Intelligence, ANSI/CTA-2089, Consumer Technology Association™, February 2020.

6. *Guidelines for Development of Civil Aircraft and Systems*, ARP4754A, SAE International, S-18, 2010.

7. *Guidelines for Conducting the Safety Assessment Process on Civil Aircraft, Systems, and Equipment*, ARP4761, SAE International, S-18, 1996.

8. Leveson, N. G., and Thomas, J. P., *System-Theoretic Process Analysis (STPA) Handbook*, March 2018. Retrieved July 13, 2022, from https://psas.scripts.mit.edu/home/get_file.php?name=STPA_handbook.pdf.

9. *Software Considerations in Airborne Systems and Equipment Certification*, RTCA DO-178, SC-145, November 18, 1981.

10. *Software Considerations in Airborne Systems and Equipment Certification*, RTCA DO-178C, SC-205, December 13, 2011.

11. *Software Tool Qualification Considerations*, RTCA DO-330, SC-205, December 13, 2011.

12. *Model-Based Development Supplement to DO-178C and DO-278A*, RTCA DO-331, SC-205, December 13, 2011.

13. *Object-Oriented Technology and Related Techniques Supplement to DO-178C and DO-278A*, RTCA DO-332, SC-205, December 13, 2011.

14. "Machine Learning, Final Report", Aerospace Vehicle Systems Institute (AVSI), AFE 87, Version 1.0, May 7, 2020, Retrieved July 13, 2022, from https://avsi.aero/projects/current-projects/cert-of-ml-systems/afe-87-machine-learning/.

15. "Artificial Intelligence in Aeronautical Systems: Statement of Concerns", SAE AIR6988, April 30, 2021. Retrieved June 26, 2021, from https://www.sae.org/standards/content/air6988/.

16. *FAA System Safety Handbook*, Chapter 3: Principles of System Safety, December 30, 2000.

17. *Standard Practice, System Safety*, Department Of Defense, MIL-STD-882E, 11 May 2012.

18. *The Army Safety, Program*, Department of the Army, US Army Regulation (AR) 385-10, 29 February 2000.

19. *Implementation Process and Tasks Supporting MIL-STD-882E With Joint Software System Safety Engineering Handbook References*, Joint Services – Software Safety Authorities (JS-SSA) Implementation Guide (IG), Rev. A, 10/10/2017.

20. *Army Military Airworthiness Certification Criteria (AMACC)*, Revision A Change 2 (C2), Prepare by: U.S. Army Combat Capabilities Development Command, Aviation & Missile Center, Redstone Arsenal, AL 35898, APR 2021.

21. *Standard Practice for Methods to Safely Bound Flight Behavior of Unmanned Aircraft Systems Containing Complex Functions*, F3269 – 17, ASTM International, September 2017.

22. Peterson, E. M., DeVore, M., Cooper, J., and Carr, G., "Run Time Assurance as an Alternate Concept to Contemporary Development Assurance Processes", NASA/CR–2020-220586, April 2020.

23. *Aviation and Missile Command (AMCOM) Software System Safety Policy, Software System Safety*, AMCOM Regulation 385-17, Headquarters, US Army Aviation and Missile Command, Redstone Arsenal, AL 35898-5000, 11 July 2020.

24. *Safety Assurance Objectives for Autonomous Systems*, V3, SCSC-153B, Safety of Autonomous Systems Working Group, Safety Critical Systems Club, Retrieved June 1, 2022, from https://scsc.uk/SCSC-153B.

25. Soudain, G., *First usable guidance for Level 1 machine learning applications*, European Union Aviation Safety Agency (EASA). Retrieved July 13, 2022, from https://www.easa.europa.eu/newsroom-and-events/news/easa-releases-its-concept-paper-first-usable-guidance-level-1-machine-0.

26. SAE G-34. (2022). "Process Standard for Development and Certification/Approval of Aeronautical Safety Related Products Implementing AI", AS6983. SAE International. DRAFT.

27. Cognilytica Cognitive Project Management for Artificial Intelligence (CPMAI), Accessed July 13, 2022, from https://www.cognilytica.com/cpmai/.

28. CRoss Industry Standard Process for Data Mining (CRISP-DM), Accessed July 13, 2022, from https://www.datascience-pm.com/crisp-dm-2/.

29. Microsoft Team Data Science Process (TDSP), Accessed July 13, 2022, from https://docs.microsoft.com/en-us/azure/architecture/data-science-process/overview.

30. Aravantinos, V., and Diehl, F., "Traceability of Deep Neural Networks", fortiss GmbH, Guerickestr. 25, 80805 Munich, GermanyarXiv:1812.06744v2 [cs.LG] 5 May 2019.

31. Amershi, S., Begel, A., Bird, C., DeLine, R., Gall, H., Kamar, E., Nagappan, N., Nushi, B., and Zimmermann, T., "Software Engineering for Machine Learning: A Case Study", 2019, Retrieved July 13, 2022, from https://www.microsoft.com/en-us/research/uploads/prod/2019/03/amershi-icse-2019_Software_Engineering_for_Machine_Learning.pdf.

32. Amodei, D., Olah, C., Steinhardt, J., Christiano, P., Schulman, J., and Mané, D., "Concrete Problems in AI Safety", , 25 Jul 2016 (v2), arXiv:1606.06565 [cs.AI], Retrieved July 13, 2022, from https://arxiv.org/abs/1606.06565.

33. "Artificial Intelligence and Machine Learning in Software as a Medical Device", U.S. Food & Drug Center For Devices & Radiological Health, January 2021, Retrieved July 13, 2022, from https://www.fda.gov/medical-devices/software-medical-device-samd/artificial-intelligence-and-machine-learning-software-medical-device.

34. Gilbert, S., Fenech, M., Hirsch, M., Upadhyay, S., Biasiucci, A., and Starlinger, J., "Algorithm Change Protocols in the Regulation of Adaptive Machine Learning–Based Medical Devices", Journal of medical Internet research vol. 23,10 e30545. 26 Oct. 2021, doi:10.2196/30545, Retrieved July 13, 2022, from https://www.ncbi.nlm.nih.gov/pmc/articles/PMC8579211/.

35. *Airworthiness of Aircraft Systems*, Army Regulation 70-62, Department of the Army Washington, DC 11 May 2016.

Safety and Product Robustness in the Air Vehicle Design Process

Barton Hainsworth
Technical Fellow
Lockheed Martin Corp
Stratford, CT. US

Ide Ehigiato
Air Vehicle Manager
Lockheed Martin Corp
Stratford, CT. US

Pete Barner
Engineering Design Manager
Lockheed Martin Corp
Stratford, CT. US

ABSTRACT

In the design of complex subsystems, Failure Modes and Effects Analysis (FMEA) exist to identify, assess, and address potential failure modes and their effects as detailed in such documents as SAE J1739 Surface Vehicle Recommended Practice, SAE ARP5580 Aerospace Recommended Practice and other published documents.

The Air Vehicle Design organization within Sikorsky, a Lockheed Martin company, utilized the aforementioned documents in particularizing the Design Failure Modes and Effects Analysis (DFMEA) for internal use by the design engineer along with select members of the integrated product development team (IPT). This internal process, while very useful, is also cumbersome and can result in such exhaustive detail such that it has not received wide adoption as originally intended due to level of effort, schedule, and budget constraints. In order to increase the safety and product robustness culture within Sikorsky down to the individual design engineer, the need exists to define a simplified process that leverages aspects of the DFMEA but serves as a tool for the engineer to refer to in the daily design decision process.

The purpose of this paper is to provide insight into this simplified approach which we refer to as the Abbreviated DFMEA (or A-DFMEA), that can be supported by Digital Transformation tools and Data Archiving as part of the digital thread. The goal is the A-DFMEA flowcharts and question sets can be contemplated by the designer for initial concept design, configuration assessments and trade studies with findings resulting in possible design modification, alternate approach determination, or mitigation/safeguard introduction. A simplified tool for use by the individual versus a formal team, can be more widely and economically utilized and begin early as possible in the product development lifecycle. The intent of the A-DFMEA is to provide for initial design corrections or progress to the comprehensive DFMEA if necessary. Though automotive in origin, the authors of this paper prefer SAE J1793 as an initial basis in developing the DFMEA and A-DFMEA processes outlined herein as a better fit for Airframe and Mechanical Systems design.

The A-DFMEA should be utilized in the initial conceptual design and trade studies and continue to mature through the completion of preliminary design. A more widespread adoption of this user-friendly tool should help foster this mindset in the design process to increase the designer's sensitivity to the ramifications of design decisions in the selection of materials, components design features, assemblies, and integration into the air vehicle. By better understanding the functional requirements and potential failure modes of a given component or system upfront, the anticipated increase in design robustness and resulting reduction or elimination of many design flaws well before they ever make their way into actual fielded systems should reap many benefits for Sikorsky and our operators. Benefits that include but not limited to: Increased safety and product design elegance/robustness, decreased cost of poor quality (COPQ), internal/external escapes and part rework. By linking the A-DFMEA with the models within the Product Lifecycle Management (PLM) system, the data archiving allows for re-use and forensics in future fielded part performance and Reliability and Maintainability (R&M) analysis.

DFMEA Overview:

The Failure Modes and Effects Analysis process is used to on the identification and mitigation of risk in the design and manufacturing of products. Properly implemented, the FMEA processes output result are both safer and more robust products, which are key elements in the rotorcraft industry. The Design Failure Modes and Effects Analysis (DFMEA) can cover mechanical, electrical as well as software systems. Additionally, the Process failure modes and effects analysis or PFMEA is related to the DFMEA in the respect to resulting manufacturing controls for the product to meet the expected design parameters must be controlled. The PFMEA, though not discussed in detail in this paper, analyzes the materials, manufacturing and assembly processes in the product development and production life, and focuses on such manufacturing operations process deficiencies to assure design performance, safety and scrap reduction. From the engineering design perspective, safety, product robustness, and performing as expected are primary concerns over scrap rate or other manufacturing considerations. The DFMEA process is very useful in identifying features or dimensions that are critical to the field performance of the system and should result in the identification of safety critical items and their controls by using the Key Characteristics (KC) on the part model/drawing for the manufacturing and quality processes. This paper focuses primarily on mechanical subsystems design and development of rotorcraft subsystems. Examples of the Sikorsky Mission Equipment subsystems include; the Rescue Hoist (with Human External Cargo) with people lifted and suspended below the helicopter and obvious single point failure points and safety considerations, Cargo Hook systems (inadvertent release concerns), Crashworthy Energy Attenuating Seats (pilot, crew/troop/passenger) with critical concerns being crash retention, occupant restraint, control of crash accelerations on the occupant, especially spinal lumbar load levels, etc.

A quote that is directly applicable to the design engineering function is "The beginning is the most important part of the work" from Plato. That seems obvious as both design approaches and decisions made during the product development process will find decades of field use with mixed results, based on those early decisions. Early identification of potential design deficiencies to eliminate by design-out through alternate approach or otherwise mitigate supports Sikorsky's objective of providing our war-fighters and commercial customers with products that meet their expectations for safety and performance in those critical missions while also reducing potential escapes, field product safety issues, and design upgrades and retrofits with those associated cost through the Engineering Change Proposal (ECP) process during the product's design operational life. Engineering escapes and resulting cost of poor business quality (COPBQ) negatively affect air vehicle reliability and safety, and ultimately company brand reputation. The objectives and uses of the DFMEA (and as relate to PFMEA) include:

- Increasing System Safety by uncovering potential failure modes that introduce hazardous conditions.
- Assessment of mission performance related effects of critical and/or undetectable failures.
- Design Mitigation of failure impact on the final product.
- Supports the verification processes such as Acceptance Test Procedures (ATP) and Maintenance check-outs and inspections.
- Assists the manufacturing engineer (M/E) in the implementation of aircraft operations sheets (AOS), manufacturing plans and Quality Assurance (Q/A) inspections, and production checks/tests of the product and process to provide a high probability of operational success.
- Providing data for development of Technical Manuals (TM) for the identification and inclusion of Notes, Cautions, and Warnings, and both special inspections and procedures for maintenance and operator support.

Sikorsky DFMEA – SAE ARP5580 or SAE J1739:

SAE ARP5580 Aerospace Recommended Practice and SAE J1739 Surface Vehicle Recommended Practice are industry standards for DFMEA. Sikorsky DFMEA procedures incorporates both standards.

According to SAE ARP5580, "Engineers have always had to consider the effects of component failures on the systems and structures that they design. However, formal methodologies for these types of analyses were not developed until the early 1960s when the obvious safety and reliability requirements of the aerospace industry began to demand them. In the late 1960s several professional societies began to publish procedures for performing a Failure Modes and Effects Analysis (FMEA). One of the earliest of these was the Society of Automotive Engineers' Aerospace Recommended Practice, ARP926, "Fault/Failure Analysis Procedure" published in 1967. In 1974 MIL-STD-1629 (ships), "Procedures for Performing a Failure Mode, Effects and Criticality Analysis" was published and, through several revisions, established the basic approach for analyzing a system. By the 1980s FMEA had become a standard part of the design process – at least in the aerospace industry. In 1988 Ford Motor Company published "Potential Failure Mode and Effects Analysis in Design (Design FMEA) and For Manufacturing and Assembly Processes (Process FMEA) Instruction Manual" which applied the methodology to manufacturing processes as well as to the design of a product. This procedure focused on the particular needs of the automobile industry and, with input from the major American automobile manufacturing companies and their suppliers, evolved into SAE Surface Vehicle Recommended Practice, J1739, issued by the SAE in 1994.

In June of 1994, then Secretary of Defense William Perry issued a memorandum titled "Specifications and Standards – A New Way of Doing Business", which directed the Department of Defense to increase their reliance on commercial products and practices. As a result of "the Perry Memo", many US military standardization documents, including MIL-STD-1629, were cited for cancellation. Around the same time the Defense Standards Improvement Council (DSIC) was chartered to oversee the standardization reform process. DSIC coordinated its position on MIL-STD-1629 with the Society of Automotive Engineers, which through its G-11, Reliability, Maintainability, Supportability and Logistics (RMSL) Division had already chartered a subcommittee to create a new FMEA procedure updating MIL-STD-1629. The subcommittee was comprised of representatives from industry, government, and academia. In response to that charter, this recommended best practice has been developed."

Why Sikorsky Airframe & Mechanical Systems prefers J1739:

SAE J1739 vs ARP5580

- **J1739 (Automotive) standard;**
  - Applies FMEA analysis methodology to the design of a product (DFMEA) as well as to manufacturing processes (PFMEA)
  - Focuses on preventing failures during design of a system, component, or part
  - Lends itself towards completing a future PFMEA analysis in support of overall
  - Manufacturing Readiness (MR) initiatives such as PPAP (Production Parts Approval Process)

- **ARP5580 (Aerospace) standard;**
  - Assesses the safety and reliability of a system qualitatively through an FMEA, as well as quantitatively with a FMECA (Failure Modes, Effects, and Criticality Analysis) for validation

- Focuses on high-level aircraft systems, and how the operation of those systems can fail
- Not directly applicable to PFMEA analysis and overall MR initiatives (PPAP)

- **Why Sikorsky Airframe & Mechanical Systems Design prefer the J1739 standard;**
  - A focus on preventing failures from occurring due to the design of a system
  - Identify failures and mitigate them through incorporation of design controls such as; KC's, risk reduction testing, or design changes
  - Supports future PFMEA analyses and overall MR initiatives Sikorsky's Production Part Approval Process (PPAP)

DFMEA Application Indicators:

In the Airframe and Mechanical Systems realm, the engineering standard work (ESW) provides guidance in the determination if an DFMEA is indicated. Such questions include:

- Is a flight safety of flight essential part or assembly affected?
- Is it a crew/personnel safety impacted such as Human External Cargo (HEC), Crash Survivability items?
- Does the component or subsystem require qualification (e.g., MIL-STD-810, RTCA DO-160 & performance testing)?
- Is the component/subsystem a Source Control (supplier development item to Sikorsky performance/requirements specification)?
- Do legacy similar systems have a history from R&M or fleet safety actions or internal engineering lessons learned database?
- Introduction on new or previously unproven technology or relatively low TRL?
- Are Commercial Off the Shelf (COTS*) items being integrated into the design?

Note *: There was an example of a (supplier qualified as a forging) COTS part in a critical HEC application that the supplier qualified as a forging but was later replaced as a casting for a producibility improvement. Though the static load rating was acceptable, fatigue life evaluation was omitted. Subsequent fatigue test failures of the parts from castings resulted in field retrofit. Take special caution in application of COTS items which the helicopter OEM may not have all the expected configuration controls.

All new system designs, new or previously unused technologies, safety or mission critical systems, or legacy similar-to designs prone to high failure rates, are all indicators of DFMEA eligibility. By asking these questions, consideration can be taken in the non-recurring engineering (NRE) estimate for proposal cost estimate capture, and during the preliminary design phase provide guidance to the IPT Lead for subsystem down-selection.

DFMEA Team vs. A-DFMEA Individual Design Engineer:

Sikorsky engineering has in place documented Engineering Procedures (E-Procs) and Engineering Standard Work (ESW) for the Airframe and Mechanical Systems functional design groups. In these processes, the Project Chief Engineer, or Air Vehicle Integrated Product Development Team (IPT) Lead shall identify subsystems of interest and form the DFMEA Teaming arrangement. The DFMEA Team should include (but not limited to) subject matter experts (SME's) from cross-functional disciplines as required: Design Engineer(s), Manufacturing Engineers, Integrated Product Team members such as

Attributes, System Safety, Human Factors, Survivability, etc., Reliability & Maintainability (R&M) Engineer and Quality Assurance Engineers etc. Oversight by the system related Technical Fellow is encouraged. In practice, the full DFMEA, while shown to be effective, can itself be reduced in effectiveness by several factors:

- Incorrect identification or missed identification of subsystems of components for DFMEA
- Errors or omissions in the identification of important performance parameters
- Budget Constraints in multiple DFMEA Teams and manpower constraints
- Schedule constraints and program priorities resulting in assumption of risk

Because there is the potential for program management in review of the full DFMEA (team level efforts) and balancing both manpower and schedule during program execution and detail design, the DFMEA process may not always be as widely adopted as design engineering would hope. Therefore, this paper contemplates flowing an abbreviated failure modes and effects analysis (A-DFMEA) process into the standard work with the thinking of the individual design engineer to act as an initial filter very early in conceptual and preliminary design. The A-DFMEA can act as a gate and guide the decision process in design features and detail approach alternative selections. A key benefit of the A-DFMEA is that it is intended as a bottoms-up process conducted by the individual design engineer as standard work in design.

The A-DFMEA Process in Design

Whereas the formal DFME is a top-down team approach, the proposed A-DFMEA is a bottoms-up approach conducted by the individual design engineer, as part of regular design duties. This is a key difference. Critical enablers such as a Lessons-Learned database, part-type failure modes and the A-DFMEA tool are described.

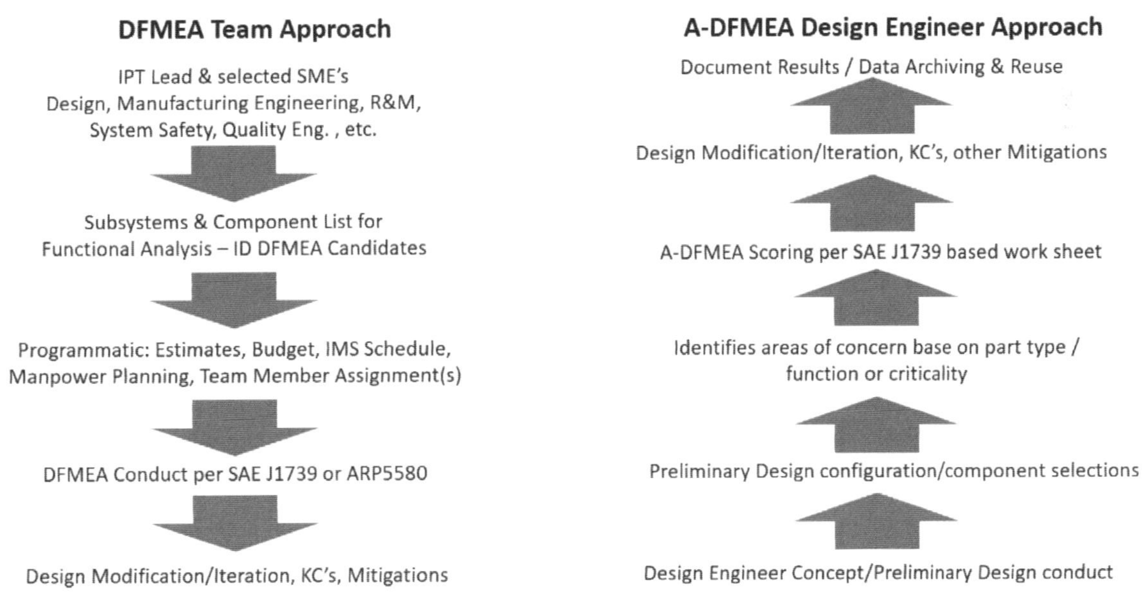

Figure 1: DFMEA Top Down vs. A-DFMEA Bottoms Up process

The driving reason behind the development of the "Abbreviated" DFMEA (A-DFMEA) process is twofold. First, as discussed earlier, a Team Level DFMEA may not capture all critical items or the effort may become target opportunities for cost reduction. Secondly, the design engineer is more intimately familiar with the design, operation and intended environment of the product, so the A-DFMEA provides and initial gate to identify potential hazards and failure modes, compare with similar products field history and group lessons learned databases by part/product type.

Potential benefits of the design engineers A-DFMEA include:

- An early assessment and gate for design approach and selection process
- Increases the Designer's awareness and more thoughtful consideration of potential failure modes related to the design approach – foster Safety Culture in the design realm
- Useful data in design trade studies and alternate approach determination
- Identification areas for application of Key Characteristic (KC's) in a design
- Reduced Design Quality Escapes
- Assists in identification of higher risk areas for mitigation through design or other barriers
- Provides documented design decision rationale for future field event investigation and lessons learned
- Can help shape qualification testing requirements for validation/verification of configuration in higher risk scenarios
- Can highlight areas of focus for targeted Reliability and Safety Engineering analysis

The design engineer should utilize the A-DFMEA questionnaire and initial score sheet to assess the design approach and its potential failure modes. Though not as exhaustive as a full DFMEA treatment, this gets the designer thinking product safety, robustness and what can potentially go wrong, how things break.

Table 1: Examples of How Products Fail

Type	Reason
Control	Electrical short
Control	Electrical Short / Sneak Circuit
Control	Erroneous drive command
Control	Inability to actuate mechanism/release
Control	Inadvertant actuation
Control	Limit Switch Failure
Control	Loss of actuator command
Control	Loss of electrical power
Control	Loss of electro-mechanical drive
Control	Loss of power ground
Control	Loss of safety device
Control	Loss of WOW input
Control	Switch/Circuit Failure
Control	Uncommanded electro-mechanical drive control
Control	Uncommanded Release
Design Envelope	Stay-Out Zone clearance loss
Design Envelope	Unforeseen Design Conditions
External	Bird Strike / Object Strike
External	External Temperature/Environment
External	Lightning Strike / Electrical Overload
External	Salt/Sand Intrusion
External	Water Intrusion
Lubrication	Contaminated Oil/Grease
Lubrication	Corrosion - lack of lubrication/grease/CPC
Lubrication	Fluid loss/leak
Lubrication	Loss of oil/lubrication
Maintenance	Incorrect Adjustment
Maintenance	Incorrect Assembly/Maintenance
Maintenance	Incorrect Torque
Maintenance	Loss/lack of adjustment
Maintenance	Poor maintenance/lubrication/cleaning
Motion	Auto-Acceleration/Deceleration failure
Motion	Bearing / roller failure
Motion	Bellcrank failure / Mount Fitting failure
Motion	Binding / Friction/Stiction
Motion	Control cable mechanism misadjusted or failure
Motion	Excessive deflection in dynamic conditions
Motion	Excessive motion outside normal operating range
Motion	Failure to extend
Motion	Failure to retract
Motion	FOD/debris in mechanism
Motion	Hinge Failure
Motion	Inability to stop
Motion	Latch Failure
Motion	Loss of brake/stop

Type	Reason
Motion	Loss of deceleration or stop limits (e.g. proximity switches)
Motion	Loss of motion -frozen joint (galling/corrosion/sand)
Motion	Mechansim stopped/jammed
Motion	Part Geometry Clash
Motion	Restricted motion inside normal operating range
Motion	Runaway condition (hoist/winch)
Part Geometry	Failure of attachment
Part Geometry	Incorrect fastener
Part Geometry	Loss of fastener
Part Geometry	Loss of Quick Release function
Part Geometry	Loss of self-locking feature
Part Geometry	Misalignment or out of tolerance condition
Part Geometry	Part Geometry/Tolerance/Fit prevents seating-locking-engagement
Part Geometry	Spring failure
Pressure	Overpressurization
Pressure	Pressure loss/leak
Pressure	Underpressurization
Structural	Buckling failure
Structural	Cracking
Structural	Dynamic Load Conditions
Structural	Excessive load condition (outside design envelope/unforseen condition)
Structural	Fatigue crack - bolt
Structural	Fatigue Induced
Structural	Fracture
Structural	Incorrect Material/Condition
Structural	Loss of structural stability
Structural	Overtorque
Structural	Reduced structural ridigity / stiffness
Structural	Rod cracking
Structural	Shock/Impact load
Structural	Static Overload
Structural	Static Overload (Bending)
Structural	Static Overload (Combined Loading)
Structural	Static Overload (Compression)
Structural	Static Overload (Tension)
Structural	Static Overload (Torsion)
Structural	Stress Corrosion Crack (SCC) Propogation
Wear/Corrosion	Adhesive disbonding
Wear/Corrosion	Chafing / Abrasion / Cut
Wear/Corrosion	Corrosion - lack of lubrication
Wear/Corrosion	Degradation/loss of energy attenuation
Wear/Corrosion	Delamination
Wear/Corrosion	Galling/Fretting/Burnelling
Wear/Corrosion	Loss of locking feature/capability
Wear/Corrosion	Webbing Restraint wear, loss of strength
Wear/Corrosion	Worn, pitted, abraded, cracked

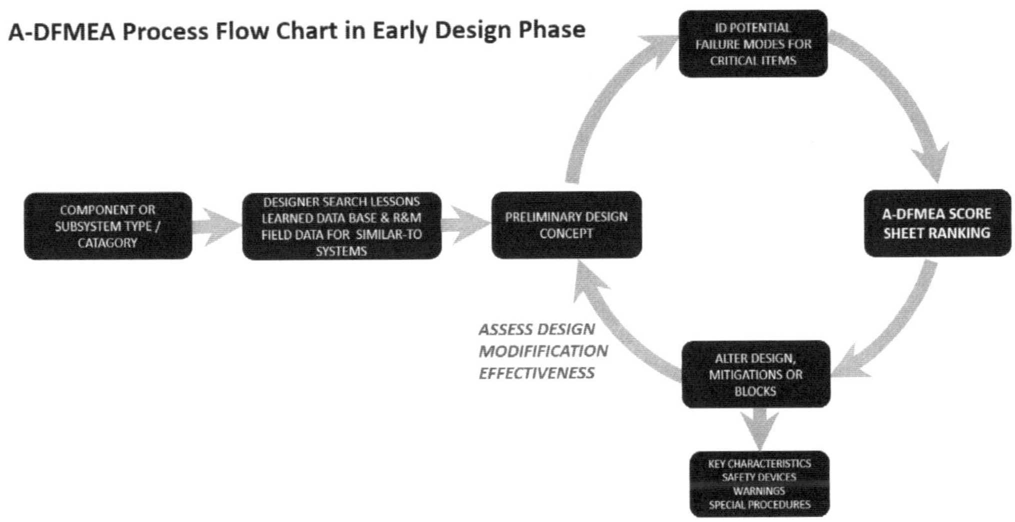

Figure 2: A-DFMEA Design Process and Iteration Loop

In the A-DFMEA process for complex systems, as part of the individual design engineer's normal duties, it is first recommended to review potential failure modes by system and components in the initial preliminary design, reviewing field history for similar type systems, and the lessons-learned database for Sikorsky Airframe, Mission Equipment/Mechanical systems for relevant systems history. The Lessons Learned capture past design, manufacture, component qualification, operational testing on aircraft, and field history such as mishap and failures, and is categorized by part family (e.g., Rescue Hoists, Energy Attenuating Seats, External Cargo Hook, Gun Mounts, etc.).

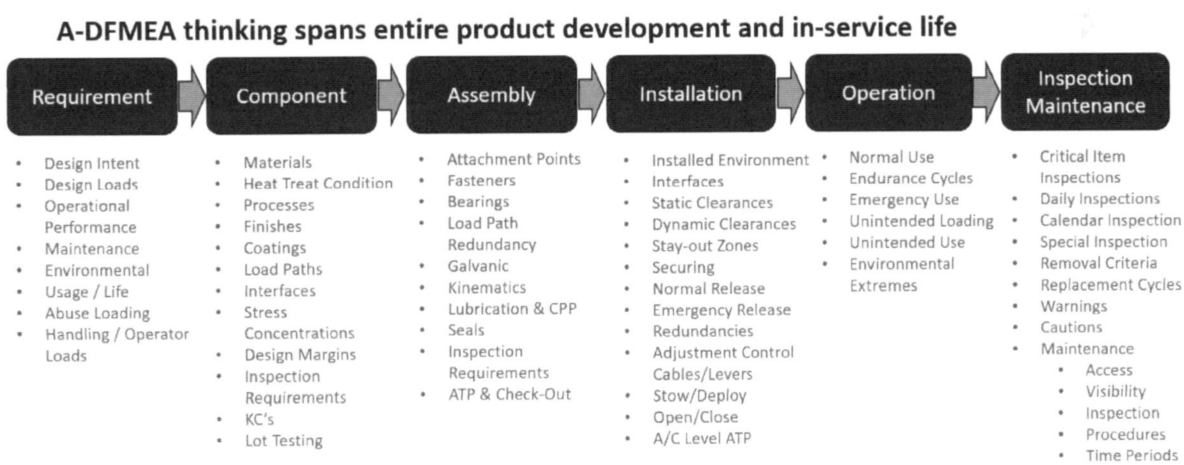

Figure 3: A-DFMEA thinking – product life cycle

When developing the design of complex systems, the design engineer needs to consider all aspects of the part manufacture, assembly installation through actual operation. An otherwise good design can easily become unreliable or introduce safety hazards if not maintained properly or misadjusted. Strive for fool-proofing the design to reduce maintainer error.

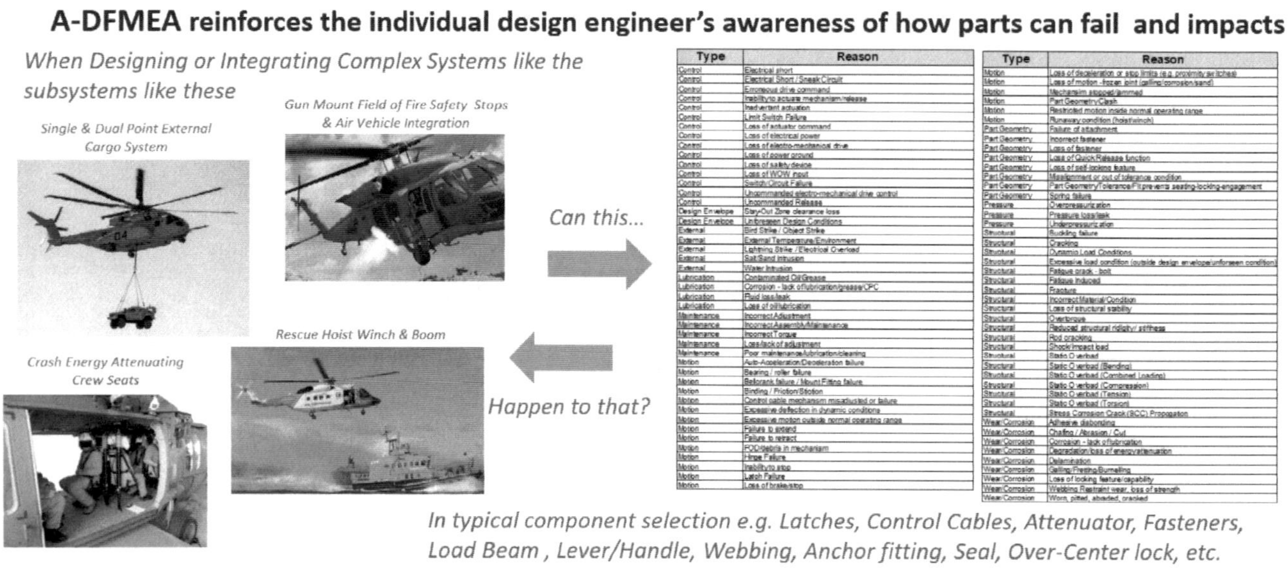

Figure 4: A-DFMEA thinking – how components fail

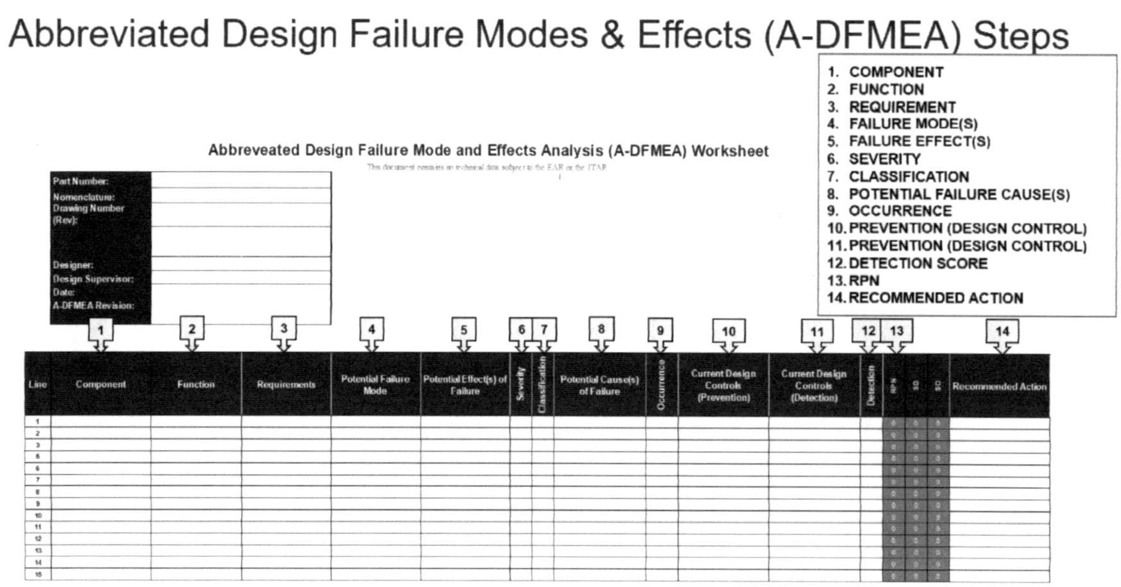

Figure 5: A-DFMEA Tool - Worksheet Process Steps

<u>How to Complete the DFMEA?</u>

Use of the Risk Priority Number (RPN) is critical in the prioritization of results.

- Rate the <u>Severity</u> of each effect of failure as defined in **Table 2**

- Rate the likelihood of <u>Occurrence</u> for each cause of failure as defined in **Table 3**
- Rate the likelihood of prior <u>Detection</u> for each cause of failure (i.e. the likelihood of detecting the problem before it reaches the end user or customer) as defined in **Table 4**
- Calculate the Risk Priority Number (RPN) by obtaining the product of the three ratings:
    - RPN = Severity x Occurrence x Detection

The RPN can then be used to compare issues within the analysis and to prioritize problems for corrective action. This risk assessment method is commonly associated with Failure Mode and Effects Analysis (FMEA).

Table 2: Defining the Severity Value

Effect	Criteria: Severity of Effect	Rank
Hazardous Without Warning	Potential failure mode affects aircraft operation and/or involves noncompliance with government regulation **without** warning	10
Hazardous With Warning	Potential failure mode affects aircraft operation and/or involves noncompliance with government regulation **with** warning	9
Very High	Aircraft/component inoperable (loss of primary function)	8
High	Aircraft/component operable but at a reduced level of performance. Customer very dissatisfied.	7
Moderate	Aircraft/component operable but comfort/convenience feature(s) inoperable. Customer dissatisfied.	6
Low	Aircraft/component operable but comfort/convenience feature(s) operable at a reduced level of performance. Customer somewhat dissatisfied.	5
Very Low	Component does not conform to fit and finish/squeak and rattle requirements. Defect noticed by most customers (greater than 75%)	4
Minor	Component does not conform to fit and finish/squeak and rattle requirements. Defect noticed by 50% of customers	3
Very Minor	Component does not conform to fit and finish/squeak and rattle requirements. Defect noticed by discriminating customers (less than 25%)	2
None	No discernible effect	1

Table 3: Defining the Occurrence Value

Likelihood of Failure	Criteria: Occurrence of Cause – DFMEA (Design Life / Reliability of Aircraft)	Rank
Very High	New technology/new design with no history	10
High (with new design, new application, or change in duty cycle/operating conditions)	Failure is inevitable	9
	Failure is likely	8
	Failure is uncertain	7
Moderate (associated with similar designs or in design simulation and testing)	Frequent failures	6
	Occasional failures	5
	Isolated failures	4
Low (associated with almost identical design or in design simulation and testing)	Isolated failures	3
	No observed failures	2
Very Low	Failure is eliminated through preventive control	1

Table 4: Defining the Detection Value

Opportunity for Detection	Criteria: Likelihood of Detection by Design Control	Rank	Likelihood of Detection
No detection opportunity	No current design control; Cannot detect or is not analyzed	10	Absolute Uncertainty
Not likely to detect at any stage	Design analysis/detection controls have a weak detection capability; Virtual Analysis (e.g. CAE, FEA, etc.) is **not correlated** to expected actual operating conditions.	9	Very Remote
After design freeze, Prior to launch	Product verification/validation after design freeze and prior to launch with **pass/fail** testing	8	Remote
	Product verification/validation after design freeze and prior to launch with test to **failure** testing	7	Very Low
	Product verification/validation after design freeze and prior to launch with **degradation** testing	6	Low
Prior to design freeze	Product validation (reliability testing, development or validation test) prior to design freeze using **pass/fail** testing	5	Moderate
	Product validation prior to design freeze using **test to failure** testing	4	Moderately High
	Product validation prior to design freeze using **degradation** testing	3	High
Virtual Analysis *Correlated*	Design analysis/detection controls have a strong detection capability. Virtual analysis (e.g. CAE, FEA, etc.) is **highly correlated** with actual and/or expected operating conditions prior to design freeze	2	Very High
Detection not applicable Failure prevention (Error-proofed)	Failure cause or failure mode cannot occur because it is **fully prevented** through design solutions	1	Almost Certain

Table 5: Key Characteristic Determination

Key Characteristic Determination
Severity Rating: 9 or 10
Severity Rating: 7-10 **AND** **Occurrence Rating: 7-10**

An Example & typical outcomes of the A-DFMEA by the Design Engineer

Presented here is a purely hypothetical of a hatch or door mechanism entry into the A-DFMEA worksheet and calculated scores based on the design engineer's identification of potential failure modes, input for the Severity, Occurrence and Detections values. The RPN is calculated from those entries. In practice, the engineer can select how detailed he or she desires to go in the analysis and can focus on more critical features of an assembly rather than an exhaustive analysis of non-critical elements. The major goal of the A-DFMEA is to increase the design engineer's awareness of the potential safety implications of ordinary design approach decisions, consideration of how things can break in real world application, intended and unintended operator use maintainer actions, in order to provide safer, more reliable products.

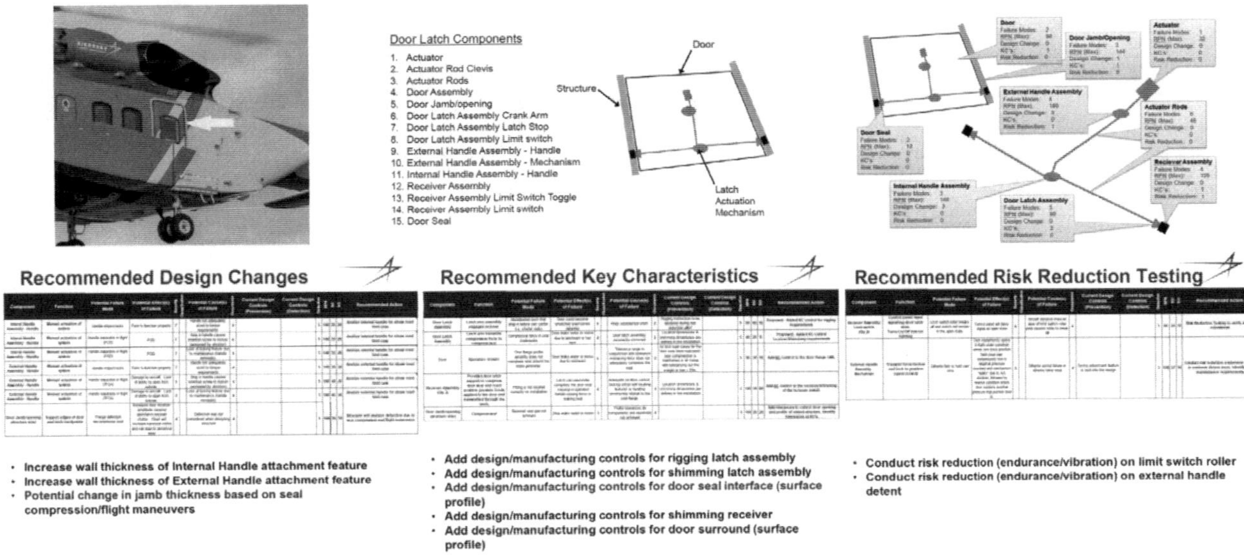

Figure 6: A-DFMEA Airframe hatch example

Recommended actions mitigate possible design failures

- Actions should be analyzed for their cost impact (non-recurring and/or recurring), and when necessary, presented to the program for funding approval

Outputs include (but are not limited to):

- Modify the design configuration or approach (Preliminary Design)
- Potential re-designs prior to release identified and implemented (Detail Design)
- Primary risks identified and mitigation plans established
- Provide focus areas for R&M and System Safety as the design matures
- Potential risk reduction test activities initiated
- Key Characteristics identified and implemented in released design
- (drawings/models)
- Modify the design, mitigate hazard level, apply KC's, Special Procedures/Inspections or Cautions in maintenance manuals.

A note on KC's for the designer: Key Characteristics in design are an attribute or physical feature whose variation has a significant influence, fit, function or performance, and are controlled on the drawing or in Model Based Definition (MBD) with special indicator to control such things as material, geometrical dimension or tolerance or performance parameter such as spring rate, etc. More details on KC's can be found in SAE AS9103.

Data Archiving & Reuse

An important element to the cultural acceptance of the A-DFMEA is the ability to find previous data and to connect the results to the designs. Sikorsky design engineers can link specific data products throughout the product life cycle to the primary design data by using the Simulation Process and Data Management (SPDM) tool as shown in **Figure 7**.

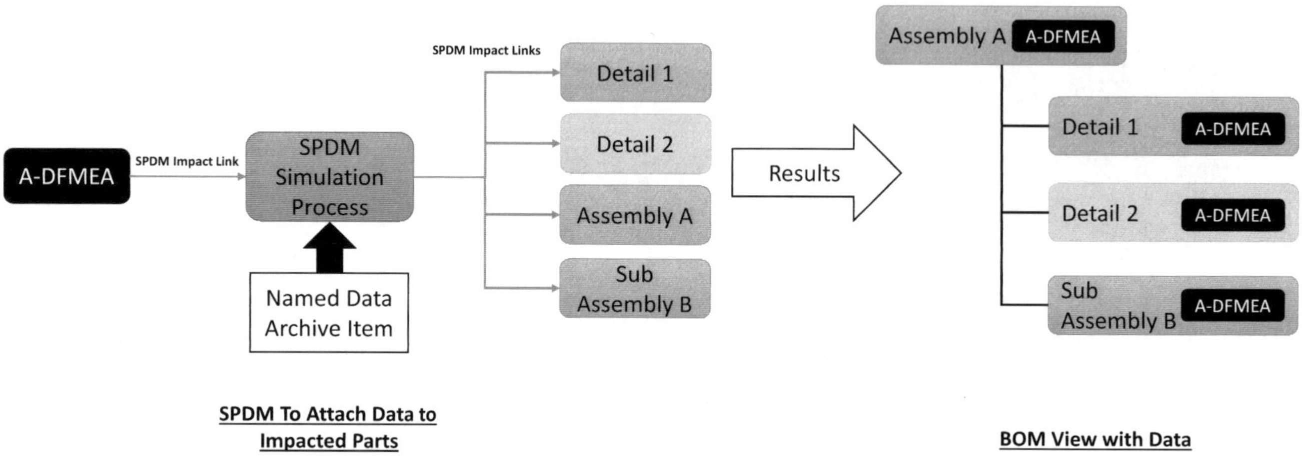

Figure 7: Attaching A-DFMEA Data with SPDM

Sikorsky's process of connecting data to designs is called Data Archiving and Reuse. The SPDM toolset allows for a lower-level A-DFMEA to roll-up to the next higher assembly, move laterally to derivative designs, and to look downwards at lower-level designs as shown in **Figure 8**.

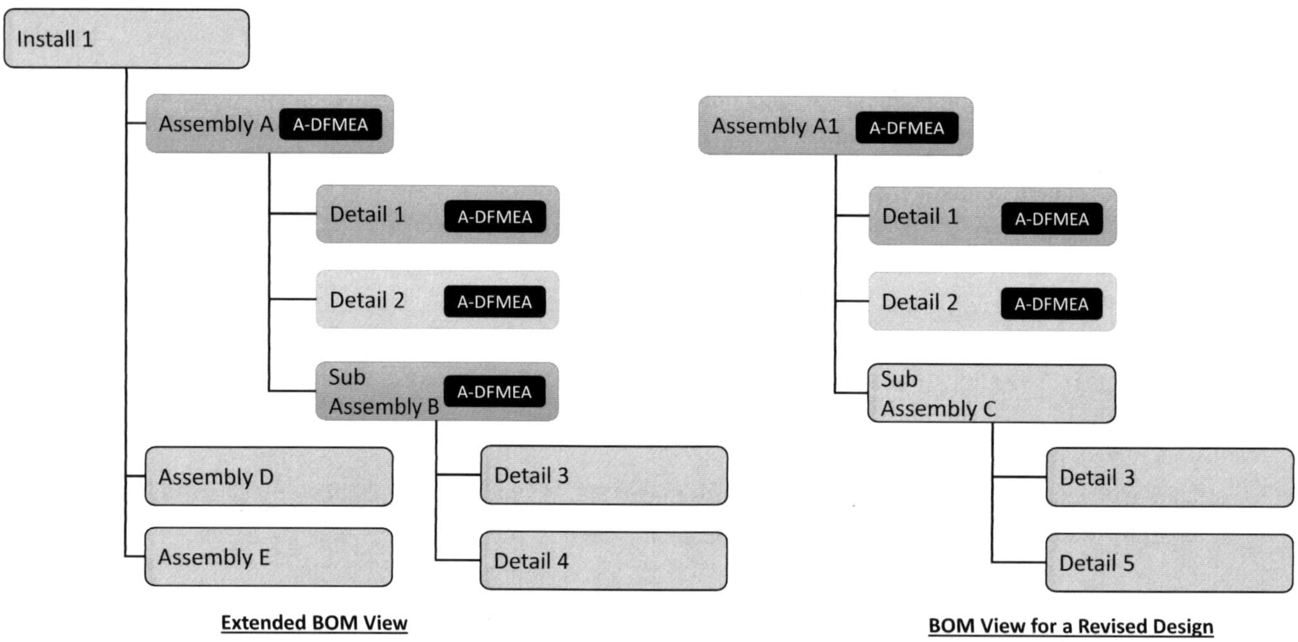

Figure 8: A-DFMEA Attached to Parts in the Bill of Material

The creation of these linkages allows for a broad impact assessment through predecessor-successor relationships as the A-DFMEA is matured when looking forward, or if a design is revised to look back at the A-DFMEA. Coupled with other Data Archiving such as trade studies, risk reduction testing, mock-up results, virtual and augmented reality evaluations, and cost estimating results, the designer has a broad array of data readily available to complete, maintain, or look back at a design. This process is currently employed on the Future Vertical Lift (FVL) projects as part of the Lockheed Martin Digital Transformation efforts. As part of the Future Vertical Lift family of aircraft, Sikorsky is engaged in a rapid prototype competition. This process enhances safety and reduces risks while enabling rapid change.

Conclusions

This study proposes a new bottoms up 'Abbreviated' DFMEA process which has the advantage of being conducted by the individual design engineer themselves and empowers them to identify, assess, mitigate or design-out potential design flaws early in the product development cycle. This process also serves to increases the engineer's focus on product reliability and safety by utilizing the A-DFMEA tool, how parts can fail sheet (modes as found in a FMECA), and especially the Lessons-Learned Database which captures by part type: Design, test, and field historical data for similar-to systems as well as field operational and safety/reliability actions. This tool will enable to designer to increase thinking about "what can go wrong" thereby increasing the product robustness and safety culture with a formalized process. The A-DFMEA shall be considered in complex systems design in-line with other parameters such as Weight, Cost and Performance. Additionally, the A-DFMEA is performed as a regular design duty as opposed to the DFMEA Team approach where only select items are identified by leadership who may or may not capture every critical subsystem, or be deferred due to manpower, schedule or budget constraints.

REFERENCES

SAE ARP5580, Recommended Failure Modes and Effects Analysis (FMEA) Practices for Non-Automobile Applications, SAE, 2020

SAE J1739, Potential Failure Mode and Effects Analysis (FMEA) Including Design FMEA, Supplemental FMEA-MSR, and Process FMEA, SAE, 2021

SAE AS9103, Aerospace Series, Quality Management Systems – Variation Management of Key Characteristics, 2012

On System Safety and Reliability Methods in Early Design Phases, Cristina Johansson, Linkoping University 2013 Thesis 1600

Modern Aircraft Flight Control Design and Bare Aircraft Sizing: an Integrated Model-Based approach

Francesco Daniele Amato
Control Law Engineer
Leonardo Helicopters Division
Cascina Costa di Samarate,
(VA) Italy

Nicola Cortigiani
Control Law Engineering Team Leader
Leonardo Helicopters Division
Cascina Costa di Samarate,
(VA) Italy

Diego Del Gobbo
Head of Flight Control System
Leonardo Helicopters Division
Cascina Costa di Samarate,
(VA) Italy

Luca Belluomini
FCS Analysis & Integration Lead
Tiltrotor Technologies
Leonardo Helicopters Division
Philadelphia,
(PA) USA

David King
Head of Tiltrotor Technologies
Leonardo Helicopters Division
Philadelphia,
(PA) USA

ABSTRACT

Emerging aircraft architectures, like Tiltrotors (Figure 1), employ advanced fly-by-wire flight control systems to reduced pilot workload, alleviate structural loads, and introduce enhanced automation and envelope protections.

The conventional rotorcraft design approach is based on a sequential Aircraft / Flight Control System iterative process that stops when an acceptable tradeoff is obtained. This approach is lengthy, often leads to a sub-optimal solution and is not suitable for the design of highly-integrated complex aircraft.

Figure 1 – Leonardo Helicopters Next-Generation Civil Tiltrotor

Modern flight control systems are highly integrated with other aircraft systems, modify aircraft dynamics and affect human-machine interface. This level of integration requires the design, implementation, and certification of a Flight Control System to be an integral part of the overall aircraft design process.

This paper proposes a multidisciplinary integrated Aircraft / Flight Control System design approach to guarantee compliance against dynamics stability requirements.

INTRODUCTION

The conventional rotorcraft design approach is based on a process addressing aeronautics disciplines such as flight mechanics, rotor dynamics and flight control theory in a sequential fashion. Design artifacts are shared within an iterative cycle until an acceptable tradeoff is obtained. This approach is lengthy and often leads to a sub-optimal design solution.

Emerging aircraft architectures, like Tiltrotors, employ advanced fly-by-wire flight control systems to reduced pilot workload, alleviate structural loads, and introduce enhanced automation and envelope protections. Multi-objective optimization plays a fundamental role in managing the growing number of design constraints and in exploiting all the flexibility a redundant fly-by-wire system can provide in these architectures. The sequential design process is not a suitable and efficient approach for the design of aircraft characterized by this level of complexity.

In this paper we address rotorcraft / Flight Control System (FCS) design in an integrated fashion. We

Presented at the VFS Development, Qualification and Affordability of Complex Systems, Huntsville, Alabama, USA Sept. 14-15, 2022.

Copyright © 2022 by the Vertical Flight Society. All rights reserved.

focus on the optimization of the FCS control moment authority and rate in conjunction with preliminary bare aircraft design, in order to meet one fundamental handling quality (HQ) requirement: dynamic stability. It is obvious that such a requirement could be satisfied by improving the stability characteristics of the bare aircraft. However, specific mission and operative requirements might exclude such a solution and stability augmentation needs to be provided by the FCS. Sizing of FCS control moment authority and rate will depend upon bare aircraft dynamics characteristics.

In the conventional design process bare rotorcraft dynamics is defined first; hence, control moment authority is selected on previous experience and tested during design iterations. Often, such iterations involve flight test activity on the prototype aircraft. As stated in the [7], up to 50% of flight testing in aerospace company is spent on improving flight characteristics based on issues discovered after initial design phase is completed.

The key challenge in addressing HQ requirements during the preliminary design phases is the lack of the physical data necessary to assemble the bare aircraft model. The strategy proposed in this work is the use of a parametric physics-based model to generate a family of linearized transfer functions, each of which is further used to calculate the required control authority and actuation speed needed to guarantee compliance against the dynamic stability requirement.

With this approach, the optimization, and the time-consuming iterations are performed to obtain the rotorcraft/FCS integration constraint specifying the required control authority and actuation speed as function of the bare aircraft dynamics. This is one of the integration constraints to fulfill in order to be compliant with handling qualities requirements.

AIRCRAFT REQUIREMENTS FLOW DOWN AND DECOMPOSITION

This section describes the transformational process (Figure 2) developed in Leonardo Helicopters (LH), and the rationale, collected during LH design experience, about enhancing the integration between the flight control and aircraft design disciplines. The transformational scheme creates a multi-disciplinary design process. The proposed approach introduces an efficient model-based design methodology that addresses the flow down of the aircraft level requirements (ALR) into system level requirements (SLR) and system integration requirements (SIR). The requirement definition process is recursive and iterative (spiral approach) and defines the required relationship between inputs and outputs, including constraints and interactions with other systems.

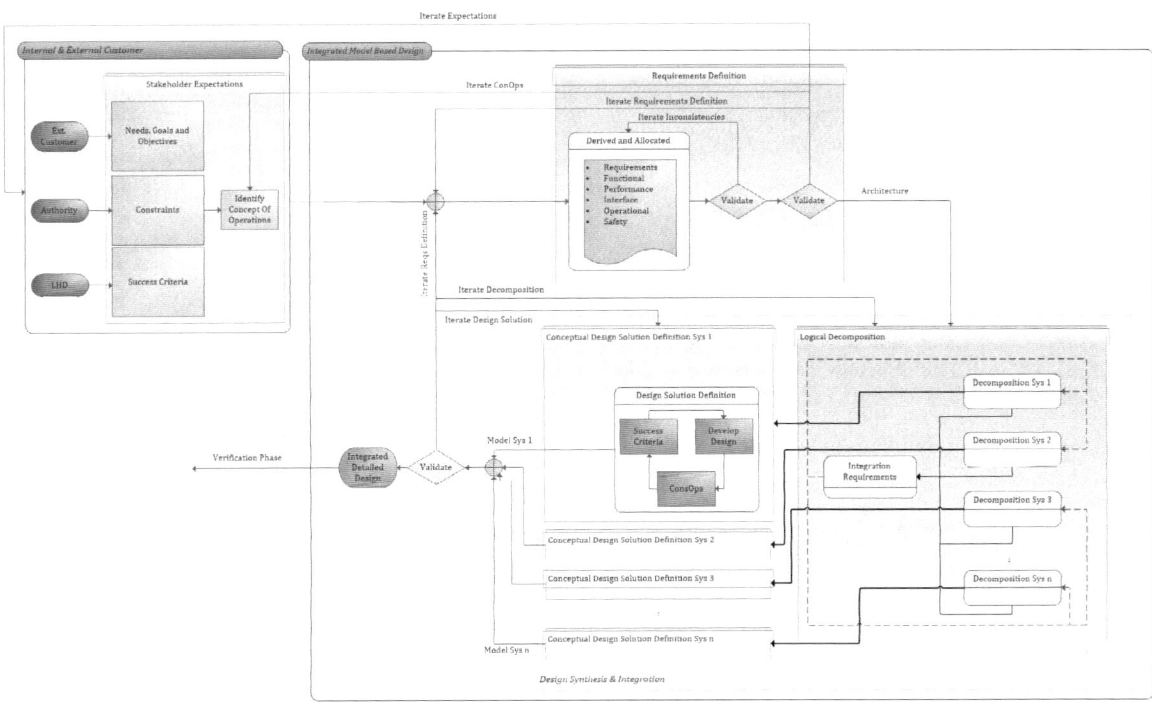

Figure 2 - Integrated System Design Process

The specification of the SLR and of the SIR starts with the definition of the design solution constraints that cannot be changed during iterations and with the identification of those elements that are under the design control of the other disciplines.

The first step of the transformational process is applied to a use-case operative scenario, which covers how the system will operate under normal condition. In this step, the interfaces between all the different disciplines and the functional expectation of the FCS are defined.

The following steps shall be then fulfilled to flow down the stakeholder expectation into system requirements and integration requirements:

- Functions definition: define what functions need to be performed to accomplish the objectives;
- Performance definition: define a quantitative measurement of how well the system needs to perform the functions;
- Reliability definition: define in a probabilistic manner the requested reliability of each function, according to the severity of any related failure scenario;
- Integration constraints definition: define the integration constraints between the different disciplines to guarantee that each system can accomplish the assigned objectives

Figure 3 is a general overview of the requirement flow down process reported in [2].

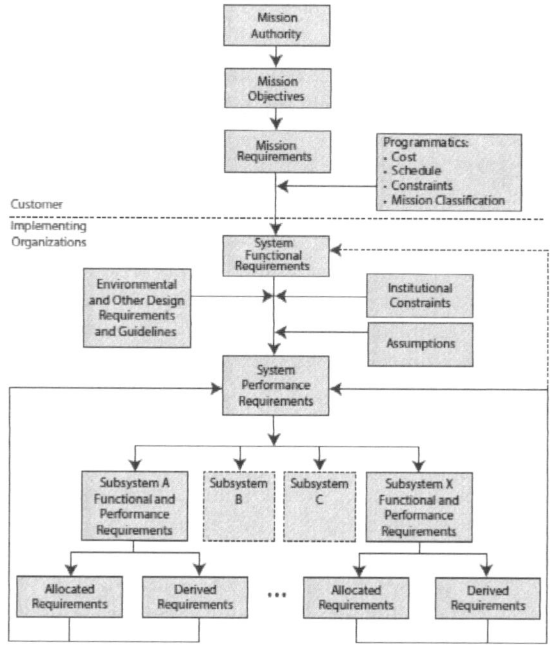

Figure 3 – Flow down requirement process

The requirements specification and validation process described above can be applied to any requirement affecting the Control Laws Design as well as the Bare Aircraft Sizing.

After system requirement validation, the detail design phase leads to the definition of a model for each system. Hence, verification against aircraft level requirements is carried out via simulation.

REQUIREMENTS VALIDATION

The core phase in a requirements definition process is the validation of the ALR against Stakeholder expectations.

The following fundamental steps shall be checked to validate a requirement:

1) Identify and correct requirements "shall" statement format errors and editorial errors;
2) Check bidirectional traceability and validate the assumption;
3) Verify that all the relevant stakeholders accept the requirement;
4) Verify that all requirements have technical sense and can be satisfied;
5) Prove that all requirements can be stated in a way that allow to verify when the solution is implemented;
6) Check that all requirements are unique and not redundant to other requirements.

USE CASE APPLICATION: DYNAMIC STABILITY

This section describes how the integrated requirement specification process described above is applied to a use-case related to dynamic stability. The goal of this use-case is to produce a quantitative requirement linking bare aircraft dynamics to control moment necessary to guarantee compliance against dynamics stability requirements.

One challenging objective in FCS design is to provide required stability and control characteristics while minimizing actuation authority and rate, to contain malfunction severity. A quantitative technique is formulated to optimize the control moment generated by the controller under the constraint specified by the dynamic stability requirements. Constraints and objectives are formulated in a manageable mathematical form to support the optimization process.

Being in the preliminary design phase, the first-guess control strategy is to use a Proportional + Integral (PI) controller. The PI acts only on the innermost loop (angular rates) and it is assumed for simplicity and generality of use. Being the simplest controller, the PI also guarantees that the results are conservative. More sophisticated controllers will be able to achieve better results during the detail design phase.

Aircraft Level Requirements

The ALRs adopted for this specific case and its mathematical formulation are the Dynamic Stability requirements reported below and extracted from the CS-29 [8] (Appendix B.VI of the CS 29).

Aircraft oscillations with period (Ts) below 20s shall be damped within a time interval that is commensurate with oscillation period, as per table below:

ID	Oscillation period [s]	Damping amplitude constraint [% of initial Oscillation Amplitude]	Damping timing constraint [s]
CS29.B.VI.a	Ts < 5	< 50%	< Ts
CS29.B.VI.a	5 < Ts < 10	< 50%	< 2xTs
CS29.B.VI.a	10 < Ts < 20	< 100%	any

Table 1 – Dynamic Stability requirement

Aircraft response with period (Ts) higher than 20s or aperiodic may present divergent characteristics with time to double constraints as per list below:

1) periodic response with Ts > 20s: minimum time to double 20s;
2) Aperiodic response: minimum time to double 9s.

Mathematical formulation of the requirements

In order to define the optimization problem, the dynamic stability requirement needs to be formulated in a mathematical form.

Consider the free motion of a linear system with state matrix A:

$$y(t) = e^{At} y(0)$$

Eq. 1

A modal decomposition of the dynamics returns the free motion of a single mode q of the system, described by the equation:

$$q(t) = e^{\lambda t} q(0)$$

Eq. 2

where λ it is associated eigenvalue.

The requirement for oscillations (Table 1) below 5s implies:

$$y(Ts) < \frac{y(0)}{2}$$

Eq. 3

(50% amplitude reduction within Ts). If this requirement is recast on individual mode we get:

$$q(Ts) < \frac{q(0)}{2}$$

Eq. 4

Thus, for the single mode:

$$q(Ts) = e^{\lambda Ts} q(0) < \frac{q(0)}{2}$$

Eq. 5

and:

$$e^{\lambda Ts} < \frac{1}{2}$$

Eq. 6

By taking the logarithm on both sides

$$\lambda Ts < \log\frac{1}{2}$$

Eq. 7

Finally, since λ can be complex, we have

$$Re(\lambda) < \log\frac{1}{2}/Ts$$

Eq. 8

The imaginary part of λ is related to the period of the oscillations: $im(\lambda) = 2\pi/Ts$. Thus, for each Ts < 5s we get an allowed region of the plane:

$$\frac{im(\lambda)}{Re(\lambda)} \leq \frac{2\pi}{\log\frac{1}{2}}$$

Eq. 9

The requirement specifies different conditions for 5s<Ts<10s and T>20s, and for unstable motion, but the approach is the same. Therefore, the allowable region (white area) for the poles in closed loop is represented in the figure 4 below:

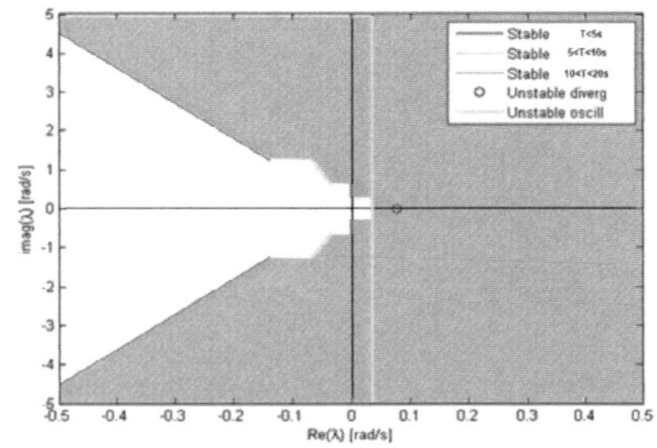

Figure 4 - Dynamic Stability Graphical Form

Following a more conservative and simple approach, the stricter requirement (blue line) is extended to the entire frequency range simplifying in this way the optimization problem.

To avoid numerical difficulties, this constraint is more conveniently rewritten as:

$$im(\lambda) - Re(\lambda)\frac{2\pi}{\log\frac{1}{2}} \leq 0$$

Eq. 10

Alternatively, this constraint can be interpreted as the minimum required damping ratio ζ of all poles: the damping ratio is defined as:

$$\zeta = \cos\theta$$

where θ, is the angle between the negative real axis and the given pole. Thus, for dynamic stability

$$\tan\theta = \frac{im(\lambda)}{Re(\lambda)} \leq \frac{2\pi}{\log\frac{1}{2}}$$

Expressing this in terms of damping ratio, the minimum required damping ratio is thus:

$$\zeta = \cos\left(\tan^{-1}\left(\frac{im(\lambda)}{Re(\lambda)}\right)\right) \geq \cos\left(\tan^{-1}\left(\frac{2\pi}{\log\frac{1}{2}}\right)\right)$$

$$\gtrsim 0.11$$

Eq. 11

This requirement could thus be replaced by an even stricter one by imposing a damping greater than 0.11 and it can be simply formulated as

$$im(\lambda) - Re(\lambda)\cos(\tan^{-1}\bar{\zeta}) \leq 0$$

Eq. 12

where $\bar{\zeta}$, is the minimum desired damping ratio.

From this point of view, it results natural to address the Dynamic Stability requirement using a pole placement framework, as detailed in the next section.

Flight Control System Requirement

A key step in the preliminary design of a Control Law is the definition of the cut-off frequency (ω_c) range of the closed loop transfer function. It can be implicitly defined as the frequency where:

$$|L(\omega_c)| = |K(\omega_c)G(\omega_c)| = 0db = 1$$

Eq. 13

where G is the transfer function of the bare helicopter dynamics, K is the transfer function of the controller and $|\quad|$ denotes the magnitude of a complex number.

Dynamic Stability requirements refer to the capability of the aircraft to reject external disturbances, like gusts and random turbulence, with specific damping characteristics.

The simplest way to achieve this objective would be to increase the closed loop cut-off frequency ω_c up to the frequency where external disturbances are less probable; unfortunately, the cut-off frequency value needs to consider possible interactions between the FCS and the airframe and rotor structural modes. Simply limiting the controller cut off frequency to below structural frequencies could result in reduced phase margins and overall reduced robustness of the controller. The integrated interdisciplinary design approach provides a methodology to tailor specific requirements to decouple the FCS to structural and rotor dynamic modes.

In order to keep as general as possible the approach, the desirable range for ω_c used during optimization is quite wide: 0.5 ÷1Hz.

To formulate this constraint mathematically, one way is to consider ω_c as a free variable within the desired range, and impose

$$|L(\omega_c)| - 1 = 0$$

Eq. 14

To impose a good robustness of the controller it is possible to require a minimum desired phase margin φ_m on each axis:

$$180 - |\angle L(\omega_c)| \geq \varphi_m$$

Eq. 15

Where $\angle$ denotes the phase operator, | | here denotes the absolute value, and the function is evaluated at the crossover frequency ω_c defined above.

Acceptable values of the minimum phase margins depend on the applicable standards and design practice as in [9].

Estimation of control moment and moment derivative

When the aircraft is subject to disturbances, the controller produces an actuation action to control the aircraft. However, this could result in a significant actuation speed request, especially if the disturbances are intense or have a significant content in the frequency range close to the cut-off frequency.

The design objective is to minimize the control moment while satisfying dynamic stability requirements. In order to have a robust design it is important to carry out the optimization considering the worst case in terms of control moment required on each axis.

The adopted approach is based on previous activity carried out in LH, aiming to identify the most critical discrete wind gust frequencies and the resulting amplitudes, which were found to be in the region between 0.2 and 0.4Hz. This assumption is based on several simulation and correlation with real flight data. The computation of the control moment on each axis and each actuator can be performed directly through the integration of the controlled dynamics subject to these discrete gusts. In particular, the ODEs of the controlled dynamics are integrated 9 times for each step of the optimization (3 wind frequencies times 3 directions). The ODEs are initially integrated for a period equal to the duration of the discrete gust (which has a 1-cosine shape), than the integration proceeds without disturbances, meaning without wind, until 20s of integration are reached.

Finally, as the aim is to minimize the maximum actuation control moment on each axes, the values of the peak absolute value of each control is computed and the following constraint is imposed:

$$u_{peak,i}(t; A_{ext}, w(t)) \leq \alpha$$

Eq. 16

Where $u_{peak,i}$ is the peak load on axis i. The peak load is functions of the full controlled dynamics A_{ext} and the wind gust $w(t)$ while α is a slack variable that acts as a moving threshold between 0 and 1.

State Space representation of PI controlled system

Consider the dynamics of a linear system:

$$\dot{x} = Ax + Bu,$$
$$y = Cx + Du$$

Eq. 17

and a set of PI controllers:

$$u = k_p e + k_i \int e\,dt = k_p e + k_i e_i$$

Eq. 18

where e_i is the state associated to the integrator of each PI controller, k_p and k_i are matrices whose sparsity pattern depends on whether the control architecture is SISO (in parallel on all axes) or fully MIMO and

$$e = y^0 - y = y^0 - Cx - Du$$

where y^0 is the set point. Thus

$$\dot{e}_i = e = y^0 - Cx - Du$$

An extended state dynamics can be written as

$$\begin{bmatrix} \dot{x} \\ \dot{e}_i \end{bmatrix} = \begin{bmatrix} A & 0 \\ -C & 0 \end{bmatrix} \begin{bmatrix} x \\ e_i \end{bmatrix} + \begin{bmatrix} B \\ -D \end{bmatrix} u + \begin{bmatrix} 0 \\ I \end{bmatrix} y^0$$

Eq. 19

Now, substituting:

$$u = k_p e + k_i e_i = k_p(y^0 - Cx - Du) + k_i e_i$$

$$(I + k_p D)u = k_p y^0 - k_p Cx + k_i e_i$$

$$u = (I + k_p D)^{-1} k_p y^0 \text{-} (I + k_p D)^{-1} k_p Cx + (I + k_p D)^{-1} k_i e_i$$

And finally

$$\begin{bmatrix} \dot{x} \\ \dot{e}_i \end{bmatrix} = \begin{bmatrix} A\text{-}B(I + k_p D)^{-1} k_p C & B(I + k_p D)^{-1} k_i \\ -C - D(I + k_p D)^{-1} k_p C & -D(I + k_p D)^{-1} k_i \end{bmatrix} \begin{bmatrix} x \\ e_i \end{bmatrix} + \begin{bmatrix} B(I + k_p D)^{-1} k_p \\ I + D(I + k_p D)^{-1} k_p \end{bmatrix} y^0$$

Eq. 20

The first term describes the dynamics of the system with the feedback PI controller, while the second term tracks the set point y^0 (if $y^0 = 0$ it gives no contribution).

The stability of the system is governed by the eigenvalues of state matrix of the first term, hence for a given bare aircraft dynamics (A, B, C, D), it depends only on k_p and k_i. Since this mathematical system is coupled, a variation in the gains on one axis affects all eigenvalues. From this consideration it is evident that the problem should be solved working with the entire dynamics of the whole system, instead of decoupling longitudinal and lateral motion as done traditionally for fixed wing. Furthermore, the lateral/longitudinal decoupling slightly perturbs the positions of the poles of the full dynamics.

Optimization problem formulation

Finally, we can formulate the multi-axis optimization problem to minimize the control moment to counteract the worst gust induced disturbance.

Mathematically, this is expressed as

$$\min_{\omega_c, k_p, k_i} \alpha$$

Eq. 21

such that:

$$\begin{cases} im(\lambda) - Re(\lambda)\cos(\tan^{-1}\bar{\zeta}) \leq 0 \; [12] \\ |L(\omega_c)| - 1 = 0\,, with\; 0{,}5 \leq \omega_c \leq 1 \;\; [14] \\ 180 - |\angle L(\omega_c)| \geq \varphi_m \; [15] \\ u(t; A_{ext}, w(t)) \leq \alpha\,, with\; 0 \leq \alpha \leq 1 [16] \end{cases}$$

Eq. 22

This is a nonconvex, nonlinear but differentiable optimization problem: eigenvalues smoothly depend on the gains k_p, k_i, $L(\omega_c)$ smoothly depends on the gains and on ω_c, and the control efforts are linear functions of a smooth dynamics. Being nonconvex, it can have multiple local optimal solutions, so either it is tackled with a global optimization algorithm, or it is tackled with a local optimization, but it is accepted that the solution found may not be the global optimum.

In order to run the optimization process a bare aircraft parametric model shall be assembled, varying the physical data within a reasonable range. In the proposed example this happen in two different steps: the first step is to identify the most significant physical parameters (e.g. vertical fin, rotor distance, etc.) that will affect the bare aircraft stability, label them and associate to them a reasonable range of variation. The second step is to choose the shape of the parametric model. This depends upon the number of parameters, the number of data points and the desired level of complexity.

The following example describes how to generate a parametric model starting from a physical model trimmed in different flight conditions, varying 4 different geometric parameters:

- Rotors Distance
- Vertical Fin Area
- Horizontal Tail Area
- Tail Rotor Diameter

The parametric model $m(x, y, z, w; p)$ returns the entire linearized dynamics matrices A, B, C, D., for each possible value of the 4 geometric parameters (x, y, z, w) in the allowed range.

Parametric modelling generates a separate model for each entry of each matrix

$$A_{i,j} = m_{A_{i,j}}(x, y, z, w; p)$$

$$B_{i,j} = m_{B_{i,j}}(x, y, z, w; p)$$

$$C_{i,j} = m_{C_{i,j}}(x, y, z, w; p)$$

$$D_{i,j} = m_{D_{i,j}}(x, y, z, w; p)$$

Eq. 23

where $A_{i,j}$ is the (i, j) entry of the reconstructed matrix A at the value of the parameters (x, y, z, w). These matrices are obtained with an interpolation within the allowed range, where p is the regression coefficients of the generic n-order model.

Below is presented an example of the bare aircraft model envelope obtained parametrizing and linearizing several physical bare aircraft models obtaining more than 10^4 dynamics models. In the figure 5 the entire envelope and the trend of the longitudinal modes (LM) and of the lateral-directional modes (LDM) is shown.

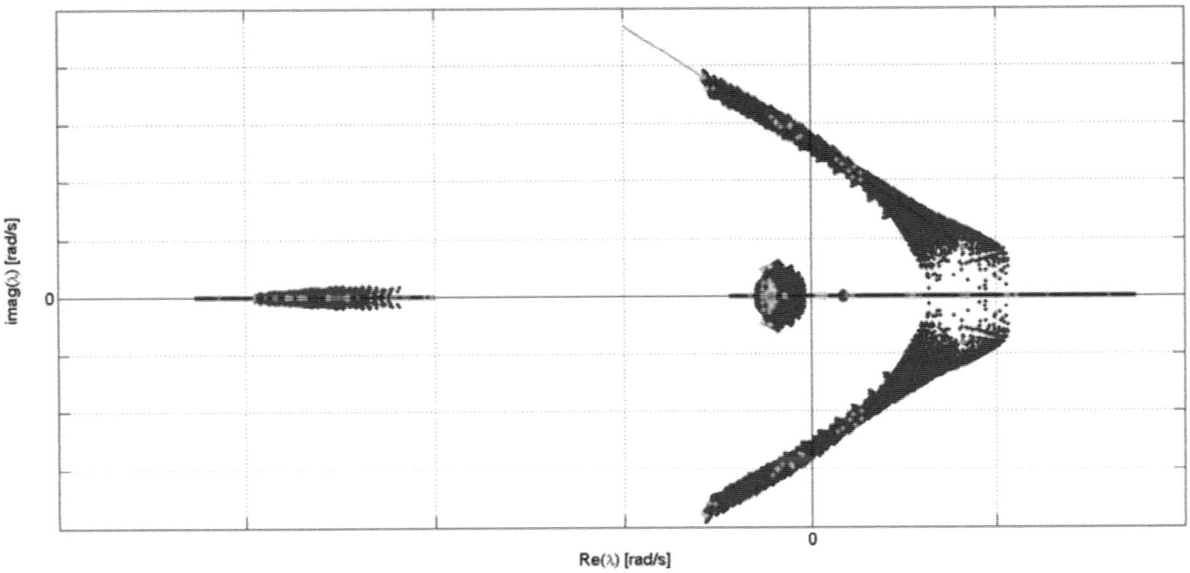

Figure 5- Poles of 21 linearized models given (green), poles of 14641 linearized models obtained (blue).

Hierarchical clustering

A valid method to aggregate all the 14641 bare rotorcraft dynamics in a straightforward representation and to properly take the highly coupled dynamics under analysis into account is represented by hierarchical clustering.

LM and LDM are preliminarily recognized considering their position in the root locus though a detailed clustering based on eigenvectors would provide more accurate results.

Once all the poles are recognised, it is possible to proceed with the clustering of the dynamical models via the clustering of the relevant poles.

Given these metrics, the aggregation of the data allows to form and identify three different clusters and to compute a bounding box for each of them, as shown in Figure 6.

This means, for example, that one cluster may be composed by dutch-roll poles within the area A and phugoid poles within any of 1, 2 or 3 area.

Clustering is performed on LM and LDM independently and three clusters are created for each axis. The Cartesian product of the clusters for the two axes covers the entire rotorcraft dynamic domain.

With three clusters for the longitudinal poles and three clusters for the lateral-directional poles, nine rotorcraft dynamics clusters are generated. Each cluster is the product of a specific cluster for LM and LDM, and thus identifies the area on the complex plane where the position of lateral-directional poles and longitudinal poles need to be to belong to that cluster. This information is very informative, as the clustering performed on the individual axis does not capture the whole aircraft stability characteristics and it neglects potential interactions arising, for example, from relatively stable phugoid poles and very unstable dutch roll poles.

The minimum values for control moment authority and rate necessary to meet the constraint [Eq.22] discussed in the previous chapters (e.g dynamic stability, gust rejection, FCS performance, etc.) are specified for each of the clusters. Furthermore, after the optimization, a Monte Carlo simulation (10000 samples) is performed with random severe turbulent gust (implemented with von Karman models) on each rotorcraft of the envelope in order to validate the results of the optimization process.

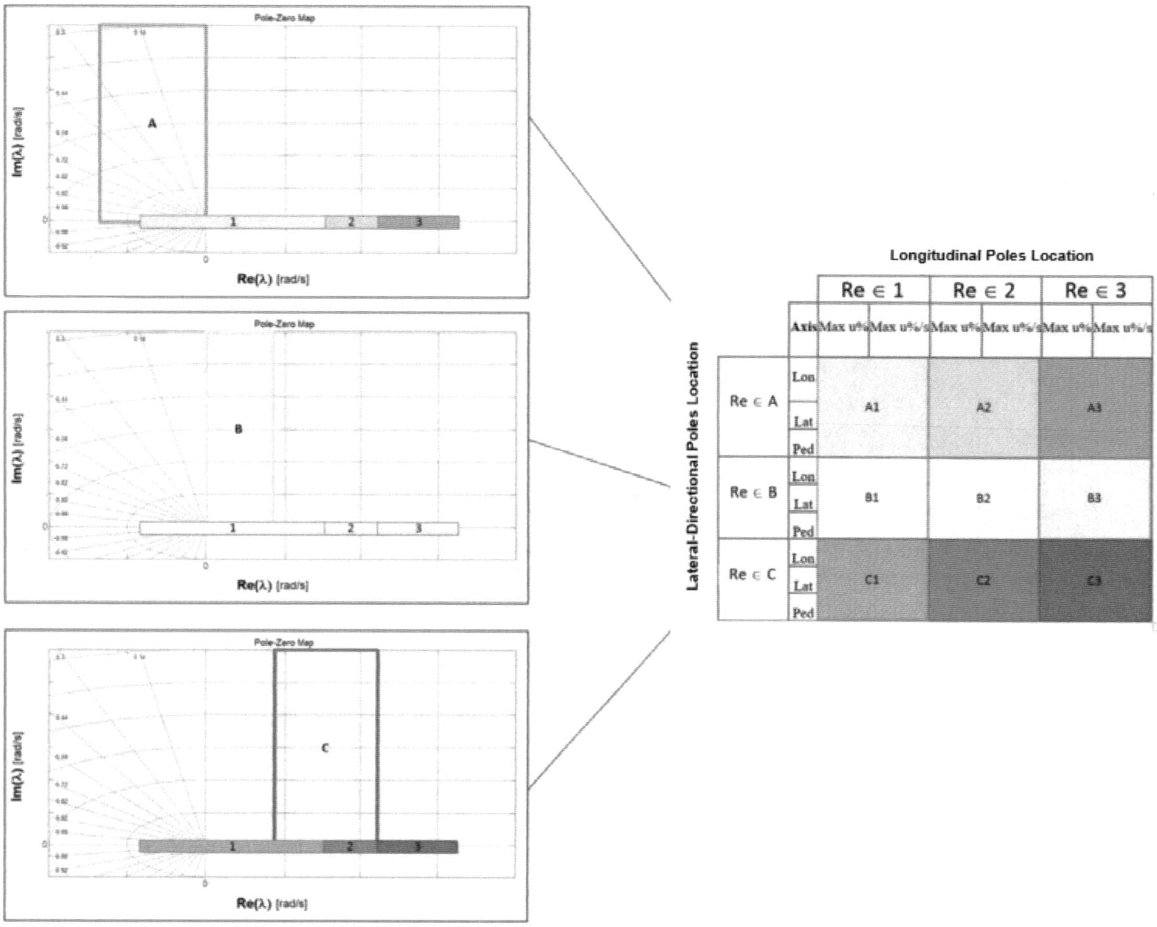

Figure 6- Example of Poles clustering representation for a generic use case

Integrated Preliminary Design Requirement

It is possible to define an integrated requirement for rotorcraft control moment sizing by solving the integrated optimization problem and considering the hierarchical clustering described above. For each linearized model reported in figure 5 and obtained from the parametric model discussed in previous section, the optimization problem [Eq.22] is solved, obtaining the minimum actuation authority and authority speed that will allow the FCS to guarantee compliance against the stability requirements [Eq.12].

This represents an integration requirement between the bare aircraft dynamics and the FCS, focusing on FCS actuation characteristics.

Specifically, the integrated requirement in Table 2 relates the flight mechanics modes, typically phugoid and dutch-roll, with the minimum control moment and control moment rate on each axis (longitudinal, lateral and directional), necessary to meet the requirements in [22].

As expected, the trend shows that increasing the aircraft instability (moving toward the right-down corner of the table), the maximum control moment and control moment rate increase (from light green to red). Since the approach proposed is preliminary and conservative, all poles with stable real part will fall in the first quadrant in table 2 (A1 quadrant in figure 6).

Lateral-Directional Pole Location (rad/s)	Longitudinal Pole Location (rad/s)	$RE \in 1$		$RE \in 2$		$RE \in 3$	
	Axis	Max u%	Max u%/s	Max u%	Max u%/s	Max u%	Max u%/s
$RE \in A$	Lon	9	21	11	21	20	40
	Lat	9	17	11	21	14	23
	Ped	6	16	7	16	7	16
$RE \in B$	Lon	9	21	12	27	21	43
	Lat	12	17	13	21	14	23
	Ped	15	16	15	17	15	17
$RE \in C$	Lon	10	21	19	27	21	43
	Lat	14	17	14	21	18	23
	Ped	17	23	24	39	24	39

Table 2 - Integrated Requirements on actuators authority and speed

Integrated detailed design verification on conventional rotorcraft

In this section we present the results of the actuation sizing process applied to two conventional rotorcraft (HC1 and HC2) configurations. The two rotorcraft actuation architectures are based on full authority trim actuation and limited authority series actuation. The series actuation provides the stability functions and the "standard sizing" adopted in FCS preliminary design phase is 10% authority and 100%/s speed. The analysis has been carried out at two different speed conditions (high speed, and low speed) and corresponding bare dynamics poles are reported in Figure 6. Results of the analysis are then compared with the standard sizing of preliminary design.

Use of integrated requirement represented by [table2] shows that HC1 and HC2 Lateral / Longitudinal dynamics fall respectively in the orange box and in the cyan box (see table 3).

Therefore, for HC1 configuration a minimum control authority of 11% with a minimum authority speed of 21%/s is sufficient to guarantee dynamics stability. Hence, the analysis confirms preliminary sizing figures of the actuation. Proper actuation sizing has been confirmed by flight test activity in severe turbulence.

For HC2 configuration a minimum control authority of 20% on longitudinal axis and 14% on lateral axis is required to guarantee dynamics stability. Hence, preliminary sizing figures of the actuation are not adequate. Flight test activity in severe turbulence with software limits on actuation has confirmed the need of the extended authority to properly reject disturbances and guarantee compliance to dynamic stability requirements.

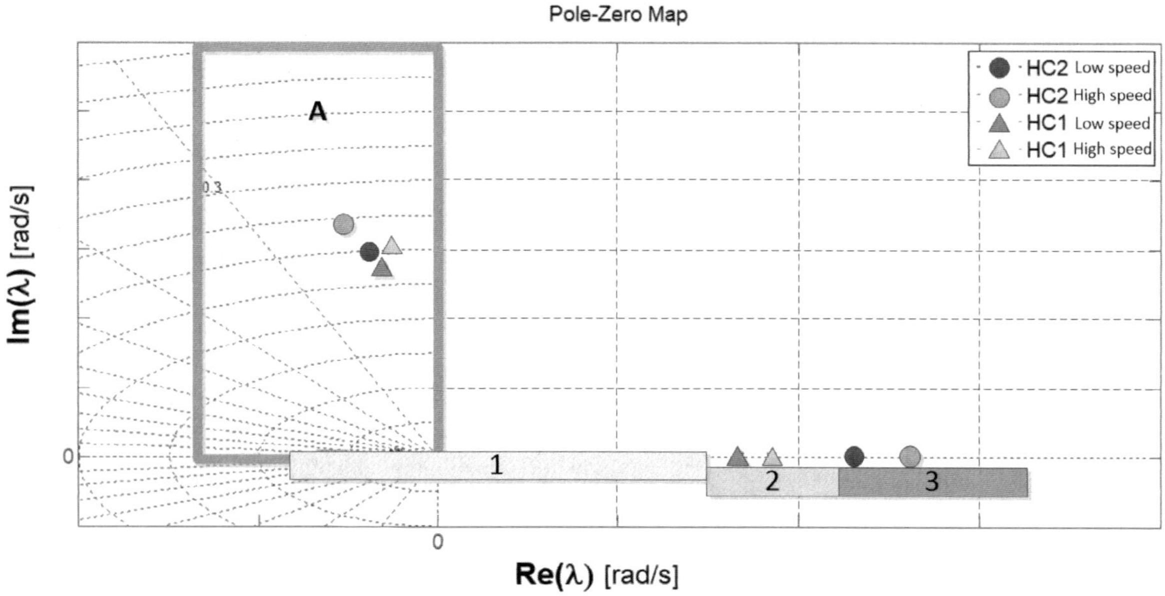

Figure 7: Root Locus of two bare rotorcrafts

Longitudinal Pole Location (rad/s)

Lateral-Directional Pole Location (rad/s)

		$RE \in 1$		$RE \in 2$		$RE \in 3$	
	Axis	Max u%	Max u%/s	Max u%	Max u%/s	Max u%	Max u%/s
$RE \in A$	Lon	9	21	11	21	20	40
	Lat	9	17	11	21	14	23
	Ped	6	16	7	16	7	16
$RE \in B$	Lon	9	21	12	27	21	43
	Lat	12	17	13	21	14	23
	Ped	15	16	15	17	15	17
$RE \in C$	Lon	10	21	19	27	21	43
	Lat	14	17	14	21	18	23
	Ped	17	23	24	39	24	39

Table 3 Comparison between actuation required for two rotorcraft configurations

Application on unconventional rotorcraft configuration

The proposed approach is general and can be adopted also for unconventional rotorcraft configurations. We have applied the methodology for sizing control moment rate in a full authority FBW FCS for a Tiltrotor configuration. Sizing objective is to guarantee compliance against dynamics stability requirements.

Figure 8 shows the envelope collecting several trimmed conditions of the bare Tiltrotor dynamics. Longitudinal poles fall in the first section (column 1 of Table 2), while lateral-directional poles fall in the second section (row B of table 2)

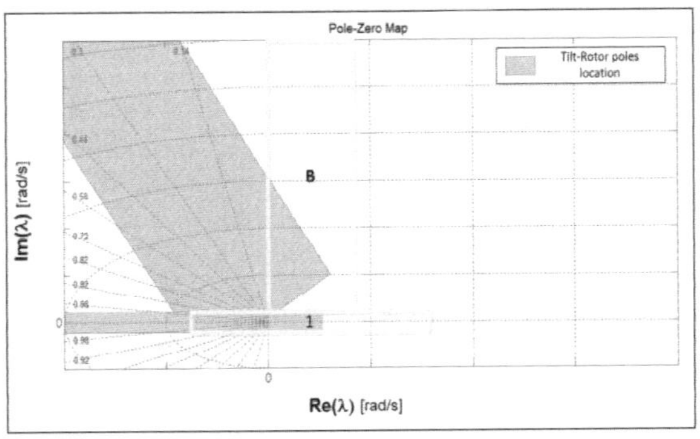

Figure 8- Tilt-Rotor Poles Map

Therefore, the control moment rate necessary to meet the dynamic stability requirements shall be greater or equal to that reported in the B1 region of figure 6; that is:

- 21%/sec for longitudinal control
- 17%/sec for lateral control
- 16%sec for directional control

This sizing only accounts for dynamics stability requirement; other requirements may lead control moment rates to larger values.

CONCLUSION AND LESSON LEARNED

The main objective of this work was to address rotorcraft / FCS design in an integrated fashion during preliminary development phase. We selected dynamics stability requirement as design objective and defined a sizing procedure for FCS control moment authority and rate to guarantee compliance against such requirement. The proposed solution is an integrated constraint that defines the minimum control moment authority and rate that the FCS shall have as a function of bare rotorcraft modes.

The procedure has been applied on two conventional rotorcraft with substantially different bare dynamics to check actuation standard sizing requirements. The analysis proved valuable in confirming actuation sizing on the first rotorcraft configuration and in modifying actuation sizing requirement in the second rotorcraft configuration. Analysis results were validated with dedicated flight test activities in severe turbulence.

The procedure has been successfully also when applied during the preliminary design of a Tilt-rotorcraft configuration, confirming the validity and broad applicability of the methodology.

These results confirm the benefits of applying a multidisciplinary integrated design approach. Such an approach avoids lengthy design iterations starting from "standard" solutions and enables development of innovative rotorcraft configurations.

Authors contact:

Francesco Daniele Amato
francescodaniele.amato@leonardo.com

Nicola Cortigiani
nicola.cortigiani@leonardo.com

Diego Del Gobbo
diego.delgobbo@leonardo.com

Luca Belluomini
Luca.Belluomini@leonardocompany.us

David W. King
David.King@leonardocompany.us

REFERENCES

1. Practical Methods for Aircraft and Rotorcraft Flight Control Design: An Optimization-Based Approach, edited by Mark B. Tischler, Tom Berger, Christina M. Ivler, Mohammadreza H. Mansur, Kenny K. Cheung and Jonathan Y. Soong, Published by AIAA American Institute of Aeronautics & Ast., 2017
2. Nasa System Engineering Handbook, edited by Shishko Robert, National Aeronautics and Space Administration, 1995.
3. Rotorcraft Flight Control Design with Alleviation of Unsteady Rotor Loads, edited by Saetti Umberto, F. Horn Joseph, Berger Tom, Tischler Mark B., published by Vertical Flight Society Forum 75At: Philadelphia, PA, 2019.
4. A Multidisciplinar Optimization Framework for Control-Configuration Integration in Aircraft Conceptual Design, edited by Ruben E. Perez, Hugh H. T. Liu, Kamran Behdinan, published by Journal of Aircraft Vol 43, No. 6, November-December 2006.
5. Conceptual Design of Tiltrotor Aircraft for Urban Air Mobility, edited by Michael Radotich, published by NASA Ames Research Center.
6. A handling qualities analysis tool for rotorcraft conceptual designs, B. Lawrence, C. R. Theodore and W. Johnson, T. Berger, published by The Aeronautical Journal, June 2018, vol 122 no. 1252.
7. Rotorcraft Handling Qualities Engineering: Managing the Tension between Safety and Performance, edited by Gareth D. Padfield, published by Journal of the American Helicopter Society, 32nd Alexander A. Nikolsky Honorary Lecture, 2013.
8. Certification Specifications and Acceptable Means of Compliance for Large Rotorcraft CS-29 Amendment 8, 24/06/2020
9. AS94900. Aerospace – Flight Control System – Design, Installation and Test of Piloted Military Aircraft, General Specification For.

Leveraging Digital Engineering to Reimagine Computational Analyses

David O'Brien
Aerospace Engineer
US Army Combat Capabilities Development Command (DEVCOM)
Aviation and Missile Center (AvMC)
Redstone Arsenal, AL, USA

ABSTRACT

Modeling and Simulation (M&S) has made substantial advances in capability and accuracy over the last fifty years. In modern design it is now routine to employ a variety of computational analyses to assess structural loads, aerodynamics, vibration, flight performance, flight controls, thermal loads, signatures, and many other technical performance areas. The focus on design has deeply entrenched M&S usage into functional engineering specialties where collaboration with other areas tends to be limited to specific data needs. This paper examines the current state-of-the-art in M&S and then looks at the state-of-the-possible. It is the author's personal belief that a new paradigm in physics-based computational analyses can be unlocked by integrating models from various functional engineering groups into a more unified model that supports the digital engineering transformation.

INTRODUCTION

Hastening the adoption of Digital Engineering (DE) processes has been a focus area for the Department of Defense (DoD) and U.S. Army over the past several years. This is emphasized by the fact that Modernization is one of the four primary Lines of Effort (LOEs) in *The Army Strategy 2018* (Ref. 1). Expanding on that strategy document, the *2019 Army Modernization Strategy* (Ref. 2) established six modernization priorities:

1. Long-Range Precision Fires
2. Next Generation Combat Vehicles
3. Future Vertical Lift
4. Army Network
5. Air and Missile Defense
6. Soldier Lethality

To ensure success, the Army Futures Command (AFC) created Cross Functional Teams (CFTs) to be enablers for each of the Army's six modernization priorities. This paper is written from an Aviation perspective so the focus will be on Future Vertical Lift (FVL), which includes the Future Attack Reconnaissance Aircraft (FARA), Future Long-Range Assault Aircraft (FLRAA), and Future Unmanned Aerial Systems (FUAS). However, DE truly has applicability to all of the Army's modernization efforts.

With multiple new systems in the modernization pipeline the Army has put a strong emphasis on modernizing its acquisition and engineering practices. Many of the Army's aviation platforms were originally designed before Computer Aided Design (CAD) was widely available. First flights of the CH-47 (1961), UH-60 (1974), and AH-64 (1975) all occurred prior to the more widespread adoption of CAD systems by industry, which didn't happen until the early 80's. Although numerous modernization efforts have taken place for those platforms since their respective first flights, there has always been legacy engineering practices to fall back to. With completely new systems in development there are no longer legacy engineering practices to fall back on and more modern DE practices need to be embraced.

In conjunction with the other strategy documents, the DoD DE Strategy (Ref. 3) was released in 2018 outlining 5 goals:

1. Formalize the development, integration, and use of models to inform enterprise and program decision making.
2. Provide an enduring, authoritative source of truth.
3. Incorporate technological innovation to improve the engineering practice.
4. Establish a supporting infrastructure and environments to perform activities, collaborate, and communicate across stakeholders.
5. Transform the culture and workforce to adopt and support digital engineering across the lifecycle.

Similarly, the U.S. Army Program Executive Office – Aviation (PEO Aviation) presented a digital transformation vision (Ref. 4) at the recent VFS Digital Enterprise

---

Presented at the VFS Development, Qualification and Affordability of Complex Systems, Huntsville, Alabama, USA Sept. 14-15, 2022. This is work of the U.S. Government and is not subject to copyright protection in the United States.
Distribution A: Approved for Public Release. PR20221354

Workshop. The common theme in all of these strategies is a desire to modernize acquisition process through leveraging Digital Engineering and transforming the workplace culture to embrace new ways of thinking.

The focus of this paper is to examine the Digital Engineering Enterprise as it pertains to the U.S. Army Combat Capabilities Development Command (DEVCOM) Aviation and Missile Center (AvMC) Systems Readiness Directorate (SRD) Aeromechanics Division's areas of expertise and seek out areas for innovation in our workflows. This paper primarily looks within the author's own organization and processes, but to be relevant must also consider how any innovative approaches fit within the greater Army Aviation Enterprise. Although this paper comes from a government perspective, most of the findings and recommendations should also have relevance to similar groups within industry.

The main motivation for this work is a reflection on the current state of computational modeling. Physics models have made great advances in utility over the years, but many times it feels like models are being applied to assess issues that have already been discovered by flight test. For example, consider the Navy's and Sikorsky's recent modeling efforts focused on the CH-53K Exhaust Gas Reingestion (EGR) problem (Ref. 5). This was an outstanding example of high-fidelity physics models being employed to identify the root cause of an issue and then analyze potential fixes. To attack this problem an aerodynamics model was integrated with a propulsion model since it was necessary to have both to study the EGR issue.

This integration of different physics models highlights the main premise of this paper. It wasn't stated in Ref. 5, but from personal practice it's known that physics models are not typically integrated together unless someone is looking for something specific. What if rather than waiting for an issue to arise, we look at model integration from the start?

GROUND RULES / ASSUMPTIONS

Being that this paper explores current processes and lays out a future vision for how the author believes we could improve, it is important to define some ground rules and assumptions.

1. This paper is purely the opinion of the author and has not been endorsed by his management or the Army. In fact, some ideas have yet to be discussed beyond first line supervisors.
2. Any mention of a helicopter Original Equipment Manufacturer (OEM) or commercial software product is in no way an endorsement of said OEM or product. They are only cited when convenient to highlight an idea or capability.
3. The primary mission of the Systems Readiness Directorate (SRD) is airworthiness. Therefore, some of the recommendations may slant towards what is best for safety as opposed to what is best for technical performance, cost, and / or schedule.
4. Much of the author's experience comes from being a member of the Aerodynamics Team of the SRD Aeromechanics Division. However, he also possesses experience in systems engineering and aerospace research, which will help to connect some of the dots beyond his present organization.

SRD AEROMECHANICS OVERVIEW

SRD has delegated airworthiness authority for most U.S. Army aviation aircraft per Army Regulation (AR) 70-62 (Ref. 6), but some other organizations also have airworthiness authority for unique circumstances. In terms of airworthiness, SRD is generally divided into two types of divisions: platform divisions and functional divisions. Platform divisions are composed of systems engineers who maintain cognizance over a particular platform and serve as the primary point of contact to the program office customers. Functional Divisions maintain functional cognizance over specific technical areas (e.g. dynamics, hydraulics, electrical, software, etc.) and are composed of a variety of engineering disciplines appropriate to the functional area of expertise. To put it simply, the functional divisions maintain the technical depth of the organization and the systems divisions maintain the technical breadth. For a complete picture of all the requirements associated with airworthiness, the reader is referred to the Army Military Airworthiness Certification Criteria (AMACC) (Ref. 7). A more descriptive overview of the airworthiness process is provided in the Rotorcraft & Aircraft Qualification Handbook (ADS-51-HDBK) (Ref. 8).

The SRD Aeromechanics Division is composed of teams that are responsible for the following: aerodynamics, dynamics, simulation, flight performance, handling qualities, flight controls, and mass properties. Each of these teams use a variety of data generated through analysis and test to make decisions regarding airworthiness in their respective areas. Due to our expertise in these disciplines, we also provide engineering support to program offices to help inform decisions with respect to various performance criteria.

A lot of SRD's work is accomplished through reviewing and analyzing engineering data provided through design reviews, analysis reports, and test reports provided from external sources (e.g. OEMs, test facilities, test ranges). However, SRD Aeromechanics Division also uses a variety of internal modeling and simulation (M&S) tools to support our assessments. The term M&S, as used here, is referring to physics modeling tools. SRD uses multiple M&S tools within each team; some are more broadly utilized while others are more for niche applications. This paper focuses primarily on a few of the more broadly utilized tools.

1. Aerodynamic analyses are performed using the Computational Research and Engineering Acquisition Tools and Environments – Air Vehicles (CREATE-AV) computational fluid dynamics (CFD) solvers Helios (Ref. 9) and Kestrel (Ref. 10).

2. Multi-body Dynamics and Aeroelastic Stability analyses are performed using the Rotorcraft Comprehensive Analysis System (RCAS) (Ref. 11).
3. Structural analyses are performed using NASTRAN (Ref. 12).
4. Flight Controls are analyzed using Flightlab (Ref. 13) and CONDUIT (Ref. 14) / Simulink (Ref. 15).

Generally speaking, each of these M&S tools are utilized by only one or two teams within the Aeromechanics Division. In certain situations, teams will share data generated by one model to provide data required by another, but for the most part the models are exercised independently. With this knowledge of this background information, we can now look at the current state of computational analysis in the division.

CURRENT STATE

In this author's opinion, the current state of computational physics-based modeling is primarily driven by functional subject matter experts (SMEs) who use tools and models that are tailored to their specific applications. Fundamentally, this isn't a wrong approach. These functional groups have evolved their tools over years of professional practice and know the short-comings and limitations of their respective analysis methods. However, communication between groups is more limited, due to the high degree of independence. This may lead to important information being unrealized or overlooked. As information is missed, the model fidelity drifts away from the aircraft truth data over time. Truth data here refences the current repository of aircraft information that is being used to design and manufacture the aircraft.

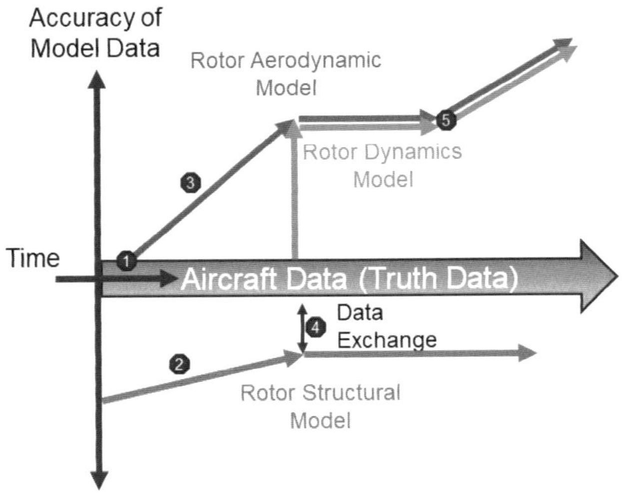

Figure 1. Representative example of the current state of computational modeling.

A representative example of how a collection of physics-based models might evolve over time is depicted in Figure 1. This example looks at the buildup of the rotor and its associated models. The large arrow in the center represents the current truth data for the aircraft. The colored lines depict different physics-based models as they evolve over time. Movement up or down of model lines on the chart represent the accuracy of the model relative to the truth data as the program evolves.

Starting at point 1, an aerodynamic model of the rotor is built from the current CAD surface or simple stacking of airfoil sections from root to tip. Typically, the aerodynamic design of the rotor comes early in preliminary design, so the aerodynamic model starts out being identical to the truth data. Meanwhile at point 2, the structural design is being refined as the loading cases are better understood, so the structural model starts moving closer to the truth model. However, the aerodynamic model may start drifting away from truth at point 3, since it isn't accounting for the structural changes to the blade.

At point 4, the structural design is defined sufficiently to generate properties for the dynamics model. Since rotor blades tend to be very flexible, it is a fairly common practice to couple the aerodynamic and dynamic models together. Once these models are coupled together, there may be a time where the error with truth stays constant. This is depicted in Fig. 1 as the period where all the model lines remain flat after the data transfer at point 4. Note that here the model lines are still depicted as offset from truth, because it assumed they are likely missing some aspect of the to be built aircraft. (All models are approximations of reality after all.) At point 5, a new aspect of the design evolves, such as refinement of the control laws, but it isn't captured by the models, so they start to diverge from truth again.

Another common source of model error relative to the aircraft truth data may come from the data being delivered at specific intervals. This is particularly true for government-based physics modeling. Government models are frequently built from data that has been transferred at specific instances in time (e.g. Contract Data Requirement List (CDRL) deliverable). As designs evolves over time, the previously delivered CDRL may never require a contractual update, which causes the accuracy to drift away from the truth data. This issue may be slightly different on the industry side, but with the use sub-contractors and geographically dispersed teams, I'd imagine that similar issues may arise.

FUTURE VISION

In the author's opinion, one of the biggest challenges with the current state is that the various model disciplines tend to be siloed and only share information at discrete intervals. A better approach is to work interdependently to achieve a more unified model. This is depicted in Figure 2. In this vision, the various modeling groups work to increase collaboration at point 1. By working together, they achieve more interoperability between models. Once this is achieved, the modelers leverage more readily available model data from modernized DE practices to ensure that the models continue to be tied to the truth data at point 2. As a result, at point 3 the physics models are able to provide high quality simulation data to inform better decisions. Points 2 and 3 are a continuous feedback loop that's maintained throughout the aircraft life cycle.

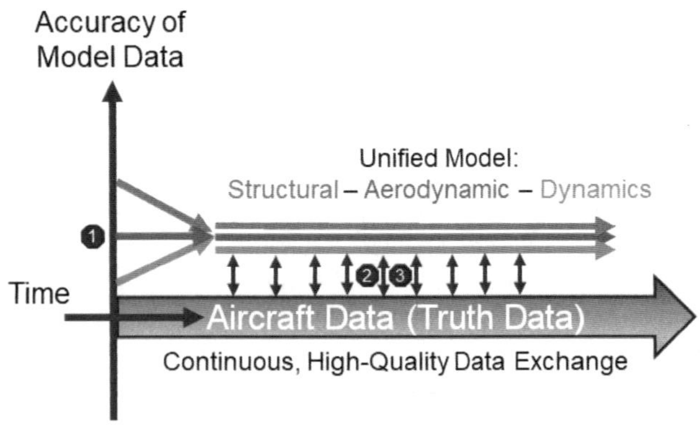

Figure 2. Future vision for computational modeling.

The key benefit of this vision is that the unified model enables the exploration of the design space for potential integration issues that won't otherwise be discovered by a single physics model. This enables true prediction of the design space. Without integrating the models together any type of physics-based model exploration is making a priori assumptions that neglected physics are negligible. It is important to note that just because a unified model is available doesn't mean it will always make sense to run the model with all physics active. Every active component of a model is going to have an associated computational cost, so one still needs to be judicious about which cases are simulated.

One interesting fall out of this vision, is that in unifying the models, a functional group is no longer the sole owner of their particular model. For example, the aerodynamics model shouldn't be changed without the buy-in from the structures and dynamics modelers. This idea may sound crazy to the functional engineer, who is used to having total ownership of their internal models. However, it's not that crazy if you think of the model in similar way as you think of the aircraft. A propulsion engineer can't move the physical location of the engine without consulting the other groups first. Why should they digitally be able to do whatever they want with the propulsion model? With respect to that thinking, the unified model proposed here is truly a digital representation of the aircraft and needs to be managed similarly to the physical aircraft.

BARRIERS

Multiple challenges potentially exist to implementing this vision, but the most prominent are listed below:

Resistance to Change – Functional groups have a lot of experience and knowledge of their disciplines. They may not see any reason to change the process that has "always worked before". Unifying the models would cause each group to lose some control over their respective models.

Fundamental Physics – Physics models have different equations and solution techniques, which can complicate communication between two models (e.g. time domain vs. frequency domain, elliptic vs. parabolic vs. hyperbolic equations, etc.).

Rethinking Verification, Validation, and Accreditation (VV&A) – Modeling and simulation requires a VV&A process per AR 5-11 (Ref. 16). Physics models are traditionally accredited by subject matter experts through an informal process. A unified model would require a more formal process for VV&A.

Cost – To carry out this vision, there will be a life-cycle cost associated with maintaining the unified physics model. Figure 2 shows a continuous cycle of the unified model being updated from truth data and providing information back.

ACHIEVING THE GOAL

Achieving the goal of building a more unified physics model requires three major lines of effort (LOE), targeting the primary barriers to success. The three LOEs are 1. Collaboration, 2. Partnering with Research & Development (R&D), and 3. Formalizing VV&A. The fourth barrier, cost, doesn't require a specific line of effort since the proposed model unification is a value proposition. It needs to demonstrate its value to justify funding it.

LOE 1. Collaboration – To successfully unify any of the wide variety of physics models, the functional SMEs need to make a deliberate effort to collaborate information more routinely. One potential solution would be to designate an integrated product team (IPT) for the unified physics model. Ideally this IPT would be led by someone with knowledge of modeling, but not necessarily biased towards any particular aspect of the physics model. In other words, you want someone that is focused on producing the informative data from the model and not just someone that is modeling for the sake of modeling. The IPT should meet regularly to discuss any planned updates to aspects of the model to give the other stakeholders a chance to express any concerns. The IPT would also give functional engineers the chance to communicate any information that might be relevant to other groups that may not have come up in their meetings. For example, a propulsion engineer might bring up changes to the exhaust duct that the aerodynamics engineer wasn't aware of.

LOE 2. Partnering with R&D – Successfully unifying different types of physics models will require working with the developers of each type of model. It's likely that not all models are readily able to communicate with other physics models due to differences in solution approach. To be successful physics models will generally need to move towards using data libraries that can be shared across various solvers. This is the approach adopted by the CREATE program (Ref. 17), whose solvers are set up as a framework. The framework approach allows component solvers to be more plug-and-play as long as they follow a set of defined interfaces and access data in the common library. Through partnering with the code development community, any issues that are discovered through the model unification process can be relayed back to the development community to enable them to build more informed development plans.

LOE 3. Formalizing VV&A – Physics models tend to be informally validated by the teams that use them. Since everyone on the team develops a working level knowledge of the models utilized, they often lack documentation. In the vision of the unified model, there is greater coupling between the digital representation of the aircraft and the physical aircraft. Therefore, the models will require more formal VV&A, which will entail better documentation of interfaces, more detailed analysis planning, and documenting results. Fundamentally, this is about trust. For the program to trust the model data that is being generated there needs to be more formal documentation defining the model capabilities and limitations.

FACTORS FOR SUCCESS

This vision for unifying various physics-based models may seem like a science project, but in actuality it's not as far-fetched as it might seem. There are numerous examples of different groups coupling a variety of tools to achieve different objectives. One of the most prominent examples, is the efforts of the Army, NASA, Bell, Boeing, Sikorsky, and academia in advancing the capability for computational fluid dynamics (CFD) and computational structural dynamics (CSD) coupling through the UH-60 Airloads Workshop (Ref. 18). The CH-53K EGR example (Ref. 5) mentioned in the introduction showed that CFD could be coupled with propulsion effects. ATA Engineering and Georgia Tech demonstrated progress towards coupling a structural model of a fuselage with a CFD solver (Ref. 19). Numerous other examples are out there, but these three are particularly relevant to Army work since they use the CREATE-AV toolsets.

Unification of tool chains is also a common theme in the industry. For example, Dassault Systèmes (Ref. 20) and ANSYS (Ref. 21) both refer to families of tools for engineering work. Their case studies show their respective tools taking you from design to manufacture. Both reference physics modeling capabilities in their families of tools, but the author does not have any experience with either to comment any further on how well those capabilities are integrated into the bigger digital engineering system of tools. However, it's clear that they see value for their customers through the use of an integrated tool chain.

CONCLUSIONS

This paper highlights the need for evolving our physics-based modeling tools to be more unified to better benefit from and support the future digital engineering environment. Current practices typically have functional engineering groups that maintain sole ownership of independent computational analysis models. Each group independently VV&A's their models through an informal process of subject matter experts signing off on the results. The proposed future vision requires unifying the models to better predict various phenomena while avoiding blind spots that arise from siloed analyses. Many of today's issues involve an interaction of two or more phenomena that can't be accurately observed using a siloed approach.

To overcome the barriers to this vision, three lines of effort (LOE) are proposed:

LOE 1. Collaboration – Functional groups need to maintain good collaboration to ensure interoperability of their models.

LOE 2. Partnering with R&D – We need to work with development teams to ensure models can work together.

LOE 3. Formalize VV&A – Unified models require more formal VV&A to ensure interoperability and to build trust with stakeholders.

Many of the building blocks needed for the success of this vision are already in the modeling community; we just need to take the next step.

ACKNOWLEDGMENTS

Many of the ideas and concepts in this paper come from numerous discussions with colleagues and mentors. It would be impossible to list every person that contributed ideas to this paper through discussions that we've had over the years, so to all of you I'd like to say thank you. However, there are a few people that I've had frequent discussions on this topic; thank you to Dr. Robert Benton, Dr. Chris Brackbill, Mr. Bill McCandless, Dr. Marty Moulton, Mr. David Quinn, Dr. Tom Thompson, and Dr. Andy Wissink for your insights.

REFERENCES

1. The Army Strategy 2018, Website: https://www.army.mil/standto/archive/2018/10/25/?from=about (accessed 9/5/2022)
2. 2019 Army Modernization Strategy, Website: https://www.army.mil/standto/archive/2019/10/17/?from=about (accessed 9/5/2022)
3. Department of Defense Digital Engineering Strategy, June 2018.
4. Davis, R. "PEO Aviation Digital Transformation Vision," Vertical Flight Society Digital Enterprise Workshop, February 23, 2022.
5. Bowles, P., Forsythe, J., Min, B., Lamb, D., Dunn, A., Siegel, E., Frydman, J., Harrington, G., Lian, C., Kube-McDowell, M., Wake, B., Powers, R., Stratton, Z., Collins, C., Spyropoulos, J., Allen, J., Jayaraman, B., and Neerarambam, S., "The CH-53K Exhaust Gas Reingestion Trade Study: Using a Tiered CFD Modeling Approach to Solving the EGR Challenges on the CH-53K

King Stallion," Paper AIAA 2021-0031, SciTech 2021, Virtual, January 11-15 & 19-21, 2021.
6. Army Regulation (AR) 70-62: Airworthiness of Aircraft Systems, 11 May 2016.
7. Army Military Airworthiness Certification Criteria (AMACC) Revision A C2, April 9, 2021, Available at: https://www.avmc.army.mil/Directorates/SRD/TechDataMgmt/ (accessed 9/5/2022).
8. Rotorcraft and Aircraft Qualification Handbook, ADS-51-HDBK, October 21, 1996, Available at: https://www.avmc.army.mil/Directorates/SRD/TechDataMgmt/ (accessed 9/5/2022)
9. Wissink, A., Jude, D., Jayaraman, B., Roget, B., Lakshminarayan, V., Sitaraman, J., Bauer, A., Forsythe, J., Trigg, R., and Peters, N., "New Capabilities in CREATE-AV Helios Version 11," AIAA SciTech Forum, Virtual Event, January 11-21, 2021.
10. McDaniel, D., "A Summary of New and Emerging Features in HPCMP CREATE-AV Kestrel," AIAA SciTech Forum, Virtual Event, January 11-21, 2021.
11. Saberi, H., Hasburn, M., Hong, J., Yeo, H., and Ormiston, R., "Overview of RCAS Capabilities, Validations, and Rotorcraft Applications," 71st Annual Forum of the American Helicopter Society, Virginia Beach, VA, May 5-7, 2015.
12. MSC NASTRAN, Product Website: https://www.mscsoftware.com/node/163 (accessed 9/5/2022)
13. Flightlab, Advanced Rotorcraft Technology (ART), Product Website: https://www.flightlab.com (accessed 9/5/2022)
14. Tischler, M., Colbourne, J., Morel, M., Biezad, D., Levine, W., Moldoveanu, V., "CONDUIT – A New Multidisciplinary Integration Environment for Flight Control Development," NASA TM-112203, USA ATCOM Technical Report 97-A-009, June 1997.
15. Simulink, Mathworks, Inc., Product Website: https://www.mathworks.com/products/simulink.html (accessed 9/5/2022)
16. Army Pamphlet 5-11: Verification, Validation, and Accreditation of Army Models and Simulations, 30 September 1999.
17. Computational Research and Engineering Acquisition Tools and Environments (CREATE) Website: https://centers.hpc.mil/CREATE/index.html (accessed 9/5/2022)
18. Yeo, H. and Ormiston, R., "UH-60A Airloads Workshop – Setting the Stage for the Rotorcraft CFD/CSD Revolution," Vertical Flight Society Forum 77, Virtual, May 10-14, 2021.
19. Reveles, N., Blade, E., Bern, N., and Smith, M., "Capability for Fully Integrated Aeroelastic Simulation of Complete Vertical Lift Configurations," AHS Forum 73, Fort Worth, TX, May 9-11, 2017.
20. Dassault Systèmes Website: https://www.3ds.com (accessed 9/5/2022)
21. Ansys, Inc. Website: https://www.ansys.com (accessed 9/5/2022)

Analyzing Mechanical Systems Using 3DX Design Tables, Kinematic Simulations, and Dynamic Simulations

Jack Faysash
jack.a.faysash@lmco.com
Mechanical Engineer Asc.
Sikorsky, a Lockheed Martin Company
San Antonio, Texas, USA

INTRODUCTION

Recent improvements in CAD and CAE software, particularly CATIA 3DEXPERIENCE (3DX), allows for an integrated and higher fidelity analysis of complex mechanical systems. The analysis of mechanical systems requires taking a large quantity of measurements throughout the systems motion, tracking component force values, and generating swept volumes. Specific improvements include the ability to use design tables to drive mechanism commands, which is especially useful for mechanisms having multiple commands that can move simultaneously. Another major improvement is the ability to run dynamic simulations within the design environment, in which the user can define input forces and track output forces along the systems motion. The analysis methods described in this abstract would be applicable to mechanical systems such as rotor controls, yaw pedal and brake modules, landing gears, turreted weapons, transmissions, doors, and flight controls surfaces.

MECHANISMS AND MECHANICAL SYSTEMS

In engineering, a mechanism is a device that transforms input forces and movement into a desired set of output forces and movement. Mechanical systems can include several mechanisms and components to achieve a desired task using engineering principles. For example, the mechanical linkages between the pedals and brake cylinders on a helicopter would be considered a mechanism, while the entire brake system as a whole would be considered a mechanical system.

Mechanisms are established by creating joints between parts within an assembly, which restrict an assembly's unwanted degrees of freedom. Joint examples include revolute joints, prismatic joints, spherical joints, point-curve joints, and come in the form of bearings, bushings, telescoping rods, cam followers, and more. CAD and CAE software's allow engineers to create and analyze mechanisms within the virtual design space. Mechanisms are created by defining joints and drivable commands until the systems degrees of freedom are equal to zero.

The proper analysis of a mechanism through the design process is critical to ensure expected functionality. This is particularly important when the mechanism plays a key role within a larger mechanical system. For example, if the brake cylinder piston requires a force of 200 lbs. to output maximum pressure, the mechanism must be designed in a way that doesn't require the pilot to exert excessive force at the pedal.

KINEMATIC SIMULATIONS

Kinematic simulations within 3DX allow the engineer to conduct detailed positional analysis of a mechanism. This is done by relating a time step to a kinematic command(s). The simulation will iterate through each defined command over a defined time period, and multiple commands can vary over the same time step. The engineer can set up the simulation to take specified measurements at each time step, such as distance between two moving components, moment arms between linkages, or angles of rotating parts. Probes can also be setup to determine position, velocity, and acceleration of each part in the mechanism. All of the data collected from the simulation can be exported to excel to use for hand calculations and plots.

For example, take a deployable weapons launcher that rotates out of the main body of an aircraft. The engineer can define the speed at which the weapons launcher rotates in and out of the aircraft. The simulation can be setup to track the angle of the weapons launcher, clearances between the launcher and the fixed components of the aircraft, as well as linear velocities and accelerations along the weapons launcher. These data points can be tracked at each time step, then export to excel to use for any required plots.

Distribution Statement A: Approved for public release; distribution is unlimited
Presented at the Vertical Flight Society (VFS) Development, Qualification, and Affordability Complex Systems Meeting, Sept. 14-15, 2022, Huntsville, AL

DYNAMIC SIMULATIONS

Historically, designers have had to jump between software's for design and analysis. With the latest version of 3DX, designers can conduct analysis within the modeling environment using dynamic simulations. Dynamic simulations allow the engineer to define input forces and track output forces within a mechanism. For example, the engineer could define the input force at the tip of the brake pedal and track the reaction forces within the pushrod and bell crank throughout the full braking sweep. The engineer can also define additional variables within the mechanism, such as component weight and inertia, spring coefficients, torque values, and bushing stiffness. Making sure the mechanism is fully defined will increase the fidelity of the results.

Force and torque values within the dynamic simulation can be defined using logic. The engineer can define a force value with respect to any parameter using a design table or formula. For example, the input force value could be defined as a constant value for the first three seconds of the simulation, and then increased linearly for the remainder of the simulation. The brake cylinder piston, for example, increases based on an equation related directly to spring constant, stroke, and pressure. This equation can be defined in the simulation to realistically track other forces within the mechanism.

MSC ADAMS has historically been the software of choice for this type of analysis. An equivalent simulation was run in both ADAMS and 3DX for tracking a hinge moment on an aircraft door. The results were within 0.07% of one another, as captured in Figure 1.

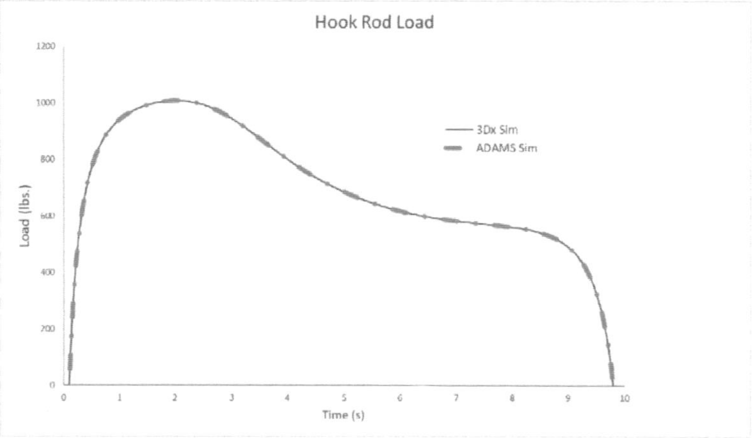

Figure 1: 3DX vs. ADAMS Dynamic Simulation

DESIGN TABLES AND EQUATION BASED PARAMETERS

3DX has a suite of tools referred to as "Knowledge Basics". With 3DX Knowledge Basics, the engineer can embed knowledge into designs such as parameters, formulas, and design/product tables and leverage them when required. Parameters are features that can be constrained by relations and can be used as the arguments of a relation, such as point or line coordinates, part geometries, and force values. Formulas are features used to define or constrain a parameter. A formula is a relation and uses operators and functions of all types whereby the engineer can perform operations on parameters. Design tables can be applied to parts or mechanical simulations and provide the engineer with a means to create and manage component families. Design tables can be used to create hundreds of models of standards parts that have unique dimensional codes, or to drive a kinematic simulation with hundreds of unique time steps. Formulas could be used to relate force values to measured parameters, such as a moment arm. Design tables and formulas are extremely efficient for automating design and analysis processes.

BRAKE MECHANISM EXAMPLE

While thus far this paper has introduced useful tools within 3DX, the remainder of this paper will focus on examples and applications of these tools. The main example this paper will dive into a Sikorsky brake linkage mechanism. The example uses a traditional, manually linked brake mechanism. The major parts consist of the brake cylinders, roller bearings, bell cranks, push rods, and the yaw pedal module. The yaw pedal module allows for rotational motion based on the pilots yaw input, and translational motion based on the pilots leg length. The brake cylinder piston can translate with respect to the brake cylinder housing, which in turn causes the pedals to rotate about their axis. For the sake of this example, the brake cylinder piston translation and force will be considered an input, while the pedal rotation and force will be considered an output or result. Figure 2 captures the motion of the yaw module and brake pedals.

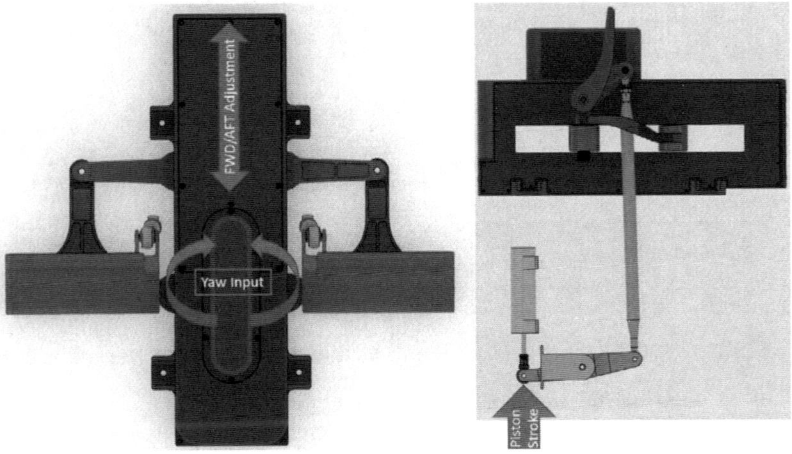

Figure 2: Yaw Module and Brake Pedal Motion

Brake systems for military aircrafts typically need to follow MIL-B-8584C requirements unless a separate specification is provided by the customer. This example will follow the requirements of a Type 1 system as defined in MIL-B-8584C, which is a manual pressure generating brake system. MIL-B-8584C requires a foot force between 15 to 20 lbs. at the tip of the pedal to cause initial movement of the brake pedal. It also requires a foot force of 75 to 125 lbs. at the tip of the pedal to develop the maximum brake pressure. In addition to force requirements, MIL-B-8584C also has pedal angle requirements in order that the pilots can maintain comfortable ergonomics. These angle requirements are captured in Figure 3. In addition to the 8584C angle requirements, standard practice is that the fully depressed pedal position never exceeds 45 degrees with the floor plane.

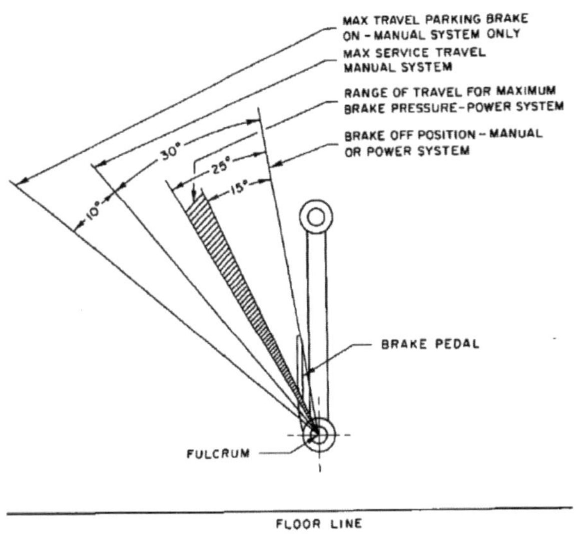

Figure 3: MIL-B-8584C Requirements

All these requirements drive the design of the brake linkage mechanism and the brake cylinder. The brake cylinder must provide enough pressure to keep the aircraft parked on an incline while the linkage mechanism balances the force of the brake cylinder piston such that the required forces at the pedal align with MIL-B-8584C requirements. The spring in the master cylinder must provide enough initial force to meet the initial pedal movement force requirement. The linkages must be designed to minimize pedal drop throughout the various yaw pedal module adjustments, which usually requires the pushrod to be as long as space permits. Taking these variables into account creates one complex problem, which is solved through an iterative design process. Now that the problem is laid out, the design and analysis process will follow the steps illustrated in Figure 4.

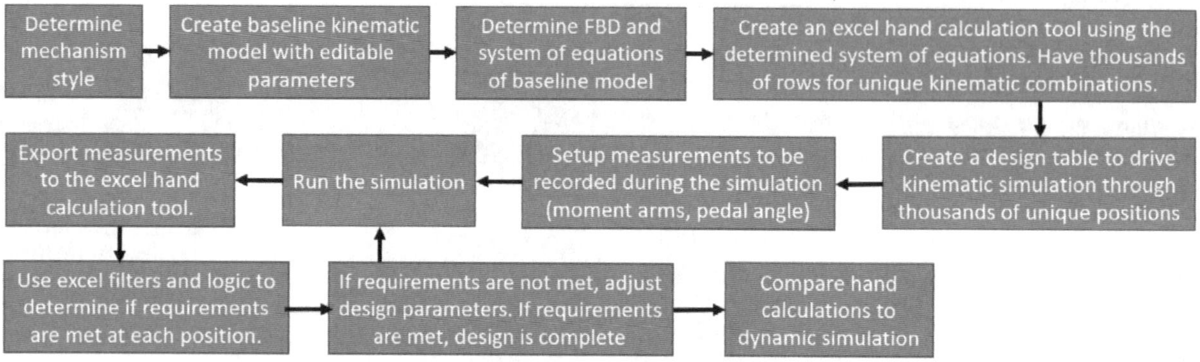

Figure 4: Brake Mechanism Design and Analysis Flowchart

The first step in the process is to determine the style of the mechanism. There are many different types of linkage mechanisms that could be used to accomplish the same goal with some being more advantageous than others. However, space and packing constraints often drive the initial concept which is iterated on by multiple design groups.

Once the mechanism style is decided, the pivot locations and linkage lengths must be further designed in a way that meets requirements. This is an iterative process of tweaking linkage parameters and brake cylinder parameters. It's best to lay out an initial concept and kinematic model to come up with a system of equations, and then go back and change parameters after running calculations for each design iteration. This is where 3DX design tables come into play for automating the design process. Dimensional parameters can be setup for each part and linked to a design table to quickly iterate through. For example, the bell crank can have a parameter that defines each pivot point. Rather than the engineer having to go back and remodel the bell crank for each design iteration, they can simply adjust a few dimensional parameters in the design table. The same concept can be used to iterate through various lengths of the pushrod.

To setup the system of equations, the engineer should first draw out free body diagrams to determine the measurements and dimensions of interest. The free body diagram and system of equations for the brake linkage mechanism are captured in Figure 5. F1 is the force the brake piston outputs, P is the pressure inside the brake cylinder, and A is the effective area the pressure acts on in the cylinder. K is the spring constant of the brake cylinder spring, L1-L4 are moment arms in the system, F2 is the force through the push rod, and F3 is the resultant force at the pedal. While the variables will hold different values for each unique combination of pilot forward and aft adjustment, yaw input, and brake cylinder stroke, the system of equations will remain the same.

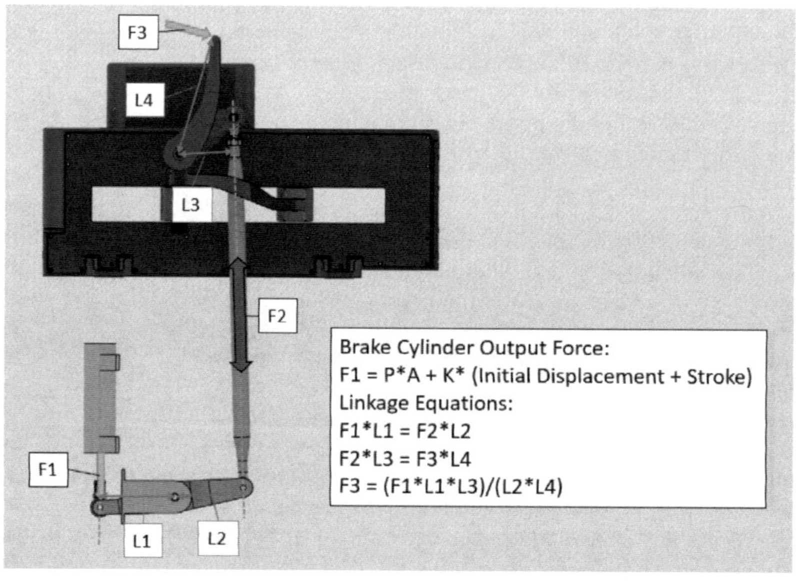

Figure 5: Brake Mechanism Free Body Diagram

Using the system of equations derived from the free body diagram, an excel calculation sheet is built to take inputs that are exported from the 3DX simulation and have columns for defined outputs. For the case of this example, moment arm lengths and pedal angles would be exported from 3DX, while brake piston force and pedal force would be calculated using the equations in

Figure 5. Filters and additional logic can be implemented for the engineer to analyze the bulk amount of data. An example excel calculation sheet is captured in Table 1.

Table 1: Brake Mechanism Calculator

Time (s)	Pilot Pedal Adjustment	Yaw	Brake Piston Stroke	L1	L2	L3	L4	Pedal Angle	Brake Piston Force	Pedal Force
1	-2	0	0	3.4	3	2.3	6	16	31	15
2	-2	0	0.5	3.5	3.1	2.6	6	27	38	21
3	-2	0	1	3.4	3.1	2.8	6	36	201	114
4	0	0	0	3.4	3.1	2.5	6	15	31	16
5	0	0	0.5	3.5	3.1	2.7	6	25	38	21
6	0	0	1	3.4	3	2.9	6	34	201	118
7	2	0	0	3.4	3.1	2.7	6	19	31	17
8	2	0	0.5	3.5	3.1	2.8	6	28	38	23
9	2	0	1	3.4	3	2.9	6	36	201	125

The next step is to create a design table within 3DX to drive the commands of the kinematic simulation over thousands of time steps. Logic can be applied using excel formulas to simulate for-loops for the drivable commands. For example, the engineer may want to loop through 100 pilot adjustment positions, with 20 yaw positions at each pilot adjustment, and 50 brake piston positions at each combination of the former two. Measurements can also be setup within the simulation to be recorded at each combination of pilot adjustment, yaw input, and brake piston stroke. Table 2 below captures an example of how an engineer might loop through kinematic commands within the simulation.

Table 2: Brake Mechanism Positions

Time (s)	FWD/AFT Adjustment (in)	Yaw Input (deg)	Right Brake Piston Stroke (in)	Left Brake Piston Stroke (in)
1	-3	-10	0	0
2	-3	-10	0.75	0.75
3	-3	-10	1.5	1.5
4	-3	10	0	0
5	-3	10	0.75	0.75
6	-3	10	1.5	1.5
7	0	-10	0	0
8	0	-10	0.75	0.75
9	0	-10	1.5	1.5
10	0	10	0	0
11	0	10	0.75	0.75
12	0	10	1.5	1.5
13	3	-10	0	0
14	3	-10	0.75	0.75
15	3	-10	1.5	1.5
16	3	10	0	0
17	3	10	0.75	0.75
18	3	10	1.5	1.5

Once the simulation is setup, the engineer can run the simulation for the baseline design and export all the measurements to the excel hand calculation tool. The engineer can review the outputs within the excel tool and determine if the design meets requirements. Adjusting design parameters multiple times before converging on a design that meets requirements is common and with the help of the design tables and automated process, the engineer saves a lot of time on performing design iterations. Once a design is converged on, the engineer would setup a dynamic simulation to compare the hand calculations to force measurements within the dynamic simulation. The engineer can define the brake cylinder piston force with a 3DX formula using the determined parameters such as piston area, spring constant, and spring displacement. Forces can then be tracked throughout the motion of the simulation, such as pushrod force and required pedal force. One downside to be mindful of is that the dynamic simulations struggle with computing multiple kinematic inputs and can be mitigated by picking a handful of pilot adjustment and yaw positions to compare with. It is also beneficial to create a swept volume of the final design to provide as a "keep out zone" to other design groups.

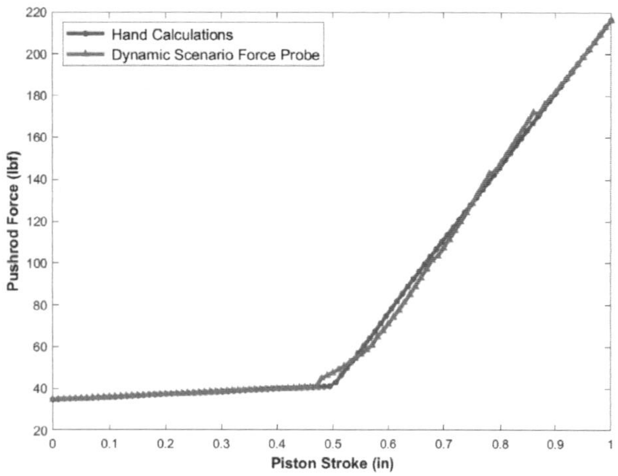

Figure 6: Hand Calculations vs. Dynamic Simulation

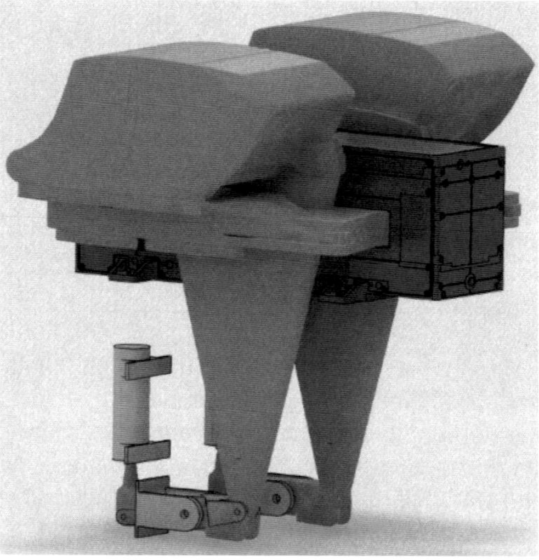

Figure 7: Brake Mechanism Swept Volume

GROUND PLANE TOOL EXAMPLE

Ground plane models are generated many times throughout the design process for various design groups to use as a constraint. For example, a radar or antenna underneath the aircraft will need to be designed in a way that prevents contact with the ground for any combination of weight and Center of Gravity (CG). Aircrafts typically have multiple design configurations, each with their own weight and CG, which in turn requires a ground plane for each configuration. It's also important to consider failure modes, such as flat strut and flat gear conditions, therefore each failure mode should also have an associated ground plane. Ground plane models must be revised throughout the design cycle. Early in conceptual design, many assumptions are made about weight and gear performance to calculate the ground plane points. As further test data is received for the landing gear shock struts and the aircraft weight configurations are matured, the ground plane models are revised to be more accurate. Later in a program's lifecycle, modifications may be made to the aircraft for new mission configurations, which will also require a regeneration of ground planes. With the help of design tables and kinematic simulations, the process of generating updated ground planes can become much more automated. The steps for creating and utilizing the ground plane tool are captured in Figure 8 below.

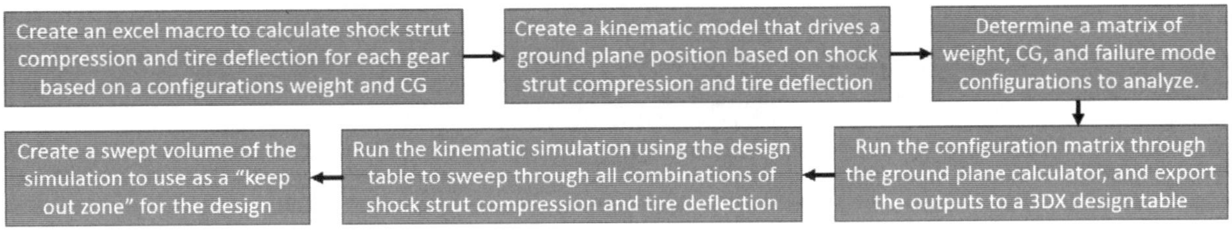

Figure 8: Ground Plane Tool Flowchart

The initial step with the ground plane tool is to gather the shock strut and tire data to create an excel macro to calculate shock strut compression and tire deflection based on weight and CG of the aircraft. As with earlier, a system of equations is created to calculate these values. Note that excel macros are not compatible with 3DX design tables, so the excel tool will need to be saved outside of 3DX.

The next step is to create a ground plane kinematic model that links a ground plane to the landing gears and is driven by shock strut compression and tire deflection. A design table is also used to drive these inputs through a series of time steps. Each time step can be thought of as a new configuration.

Once the kinematic model is created, the engineer must determine the various weight and CG configurations of the aircraft. The most critical weight will be the max design gross weight, as it will place the underside of the aircraft closest to the ground. Flat strut and flat tire conditions shall also be considered as they will place the underside of the aircraft even closer to the ground. Once all of the applicable configurations are determined, the associated weights and CG's shall be processed through the ground plane excel calculator, which will output the shock strut compression and tire deflection for each case. These outputs can then be copied to the 3DX design table to drive the kinematic simulation. Each iteration in the simulation can be viewed for its associated ground plane model, and a swept volume can also be generated to account for the worst case ground plane surface to provide to the design groups. Using these tools has proven to be beneficial in automating the ground plane modeling process and will continue to provide time savings for years to come. Figures 9 through 11 below capture examples of how the tool can be used. The majority of the landing gear geometry was hidden for release.

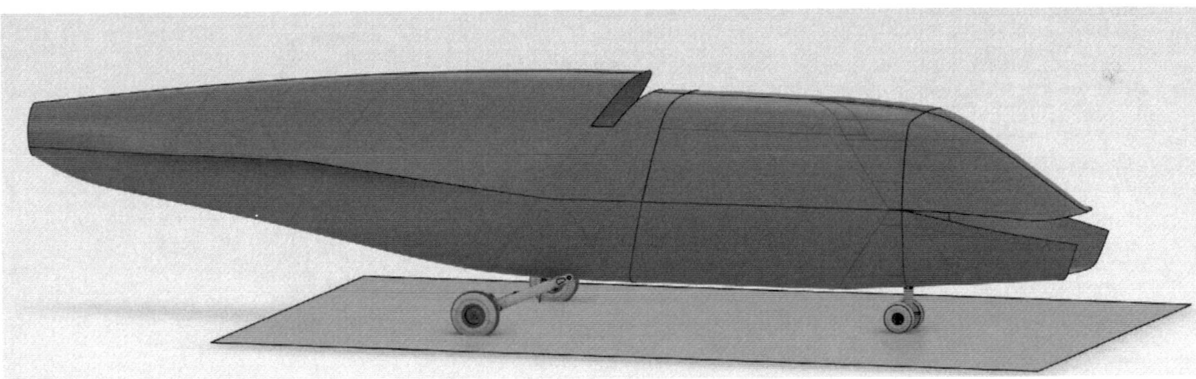

Figure 9: Baseline Ground Plane

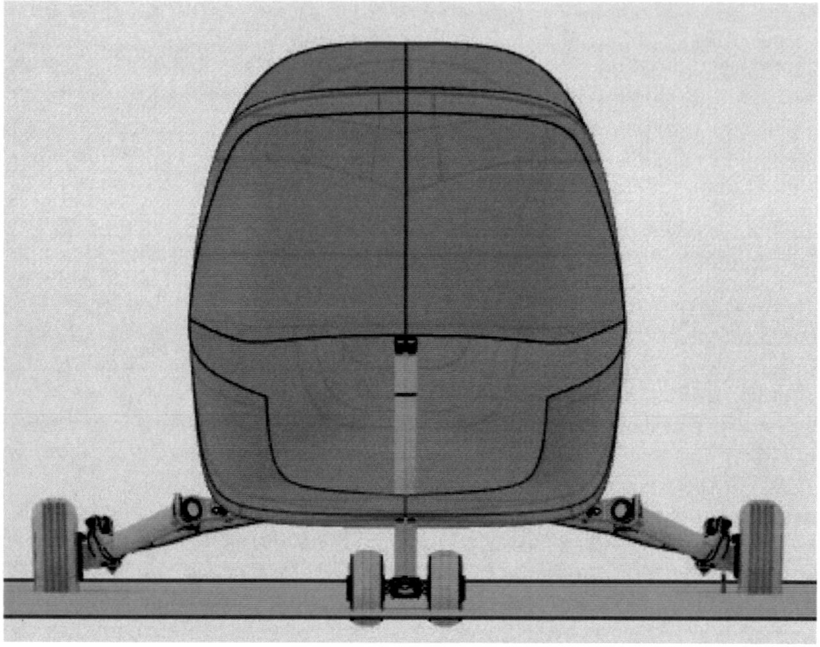

Figure 10: Flat Strut Condition

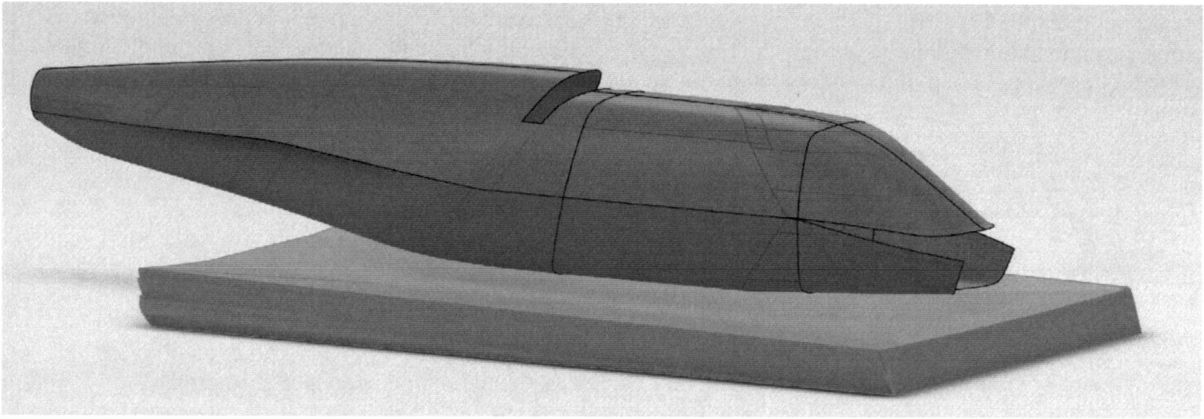

Figure 11: Ground Plane Sweep

CONCLUSION

The design and analysis methods captured in this paper will continue to be used and improved upon by Sikorsky. The method of using design tables for design and analysis will lighten the engineer's workload by making repetitive processes much more automated, while also promoting a higher level of fidelity when compared to historical analysis methods. The digital thread also allows the simulation files to be saved and linked to the overall aircraft model and can be reviewed to compare results as the aircraft advances through built and test phases. Multiple simulation models can also be combined into a single model to create marketable animations, such as simulating the aircraft through a landing and combining the landing gear and rotor simulations into one file. As digital transformation continues to charge through the aerospace industry, modeling, simulation, and analysis methods will improve and become more representative of the physical products performance.

ACKNOWLEDGEMENTS

This research was partially funded by the Government under Agreement No. W911W6-19-9-0005. The U.S. Government is authorized to reproduce and distribute reprints for Government purposes notwithstanding any copyright notation thereon. The views and conclusions contained in this document are those of the authors and should not be interpreted as representing the official policies, either expressed or implied, of the Technology Development Directorate, or the U.S. Government.

© 2022 Lockheed Martin Corporation.

Distribution Statement A: Approved for public release; distribution is unlimited
Presented at the Vertical Flight Society (VFS) Development, Qualification, and Affordability Complex Systems Meeting, Sept. 14-15, 2022, Huntsville, AL

Multidisciplinary Design of a Complex Tiltrotor Morphing Wing System

Dr. Damian Rogers
Sr. Systems Engineer, Design Architect
Leonardo Helicopter Division
Philadelphia, PA, USA

Claudio Punzi
NGCTR Structural Dynamics Chief Project
Leonardo Helicopter Division
Cascina Costa di Samarate, Italy

Alberto Angelo Trezzini
Digital Transformation Project Lead
Leonardo Helicopter Division
Cascina Costa di Samarate, Italy

Pierre Abdel Nour
Chief Designer NGCTR
Leonardo Helicopter Division
Cascina Costa di Samarate, Italy

ABSTRACT

While aeromechanics and system element interactions for typical fixed- and rotary-wing aircraft are historically well defined, the complex aerodynamic interactions for tiltrotor aircraft introduces multiple added levels of complexity. This added complexity of an aircraft that operates with the properties of both a vertical takeoff and landing (VTOL) rotary-wing and a fixed-wing aircraft adds dimensions of analysis that are not typically found in other aircraft. Unique complexities arise due to architecture where a wing, used for lift in fast forward flight configuration, is located below a rotor, being used for lift in a VTOL configuration[1]. Here, the rotor downwash creates a lift-opposing force on the wing, termed 'download'[2]. As such, model-based design and engineering approaches with parallel simulation and system design phases are required, in conjunction with a multi-disciplinary team, in order to solve the complex interactions during the design phases.

Given the issue of download on the wings due to rotor downwash reducing fuel-efficiency during VTOL operations, a need to minimize this download arose. This necessitated the design of a system that would lower the download in VTOL and hover operations, while keeping the aerodynamic wing properties in airplane mode; a "morphing wing system" (MoWS). The basis of the morphing wing was to design a wing element which could articulate out of the rotor downwash in VTOL mode and articulate up into a standard wing configuration for aircraft mode. While these goals were unique to tiltrotor aircraft, they were in line with the European Union (EU) Clean Sky 2 (CS2) initiative to reduce CO_2 emissions by 20-30% by increasing aircraft fuel efficiency[3]. Since the least fuel-efficient phases of tiltrotor operations are during VTOL and hover, increasing efficiency in these phases, through the design and implementation of the morphing wing system, is shown to reduce overall high-speed VTOL (HSVTOL) operational CO_2 emissions[4].

This paper presents the design of a morphing wing element applied to the Next Generation Civil Tiltrotor – Technology Demonstrator (NGCTR-TD), using a multi-disciplinary design team, in conjunction with the parallel use of a set of simulation models during the design phases. Given the complex interactions of the design, each model in the set took into account different aspects of the design and were integrated together to give a pseudo-live representation of the current design state, as it progressed through the iterative design loops. This led to a rapid design phase, culminating in the successful design of a morphing wing system that could reduce the download on the wing during VTOL and hover operations of a tiltrotor aircraft.

Presented at the VFS Development, Qualification and Affordability of Complex Systems, Huntsville, Alabama, USA Sept. 14-15, 2022. Copyright © 2022 by the Vertical Flight Society. All rights reserved.

NOTATION

AAM	Advanced Air Mobility
CDR	Critical Design Review
CPE	Chief Project Engineer
CS2	Clean Sky 2
EU	European Union
FCS	Flight Control System
FVL	Future Vertical Lift
HSVTOL	High-Speed Vertical Take-Off and Landing
JU	Joint Undertaking
LHD	Leonardo Helicopters Division
MoWS	Morphing Wing System
NGCTR-TD	Next Generation Civil Tiltrotor – Technology Demonstrator
PDR	Preliminary Design Review
SE	Systems Engineer/ing
STOL	Short Take-Off and Landing
UAM	Urban Air Mobility
VTOL	Vertical Take-Off and Landing

INTRODUCTION

From the inception of tiltrotor aircraft decades ago, and up to present with new tiltrotor and VTOL designs, aircraft designers have found many advantages of an aircraft that can articulate its propulsion system(s) from pointing vertically for VTOL and hover, to horizontal for fast forward flight. This, however, usually comes at a cost when the wings, which create the lift in fast forward flight modes, are situated directly below the vertically oriented propulsion system(s). Here, the wing inhibits rotor outflow, reducing effective thrust, and therefore aircraft efficiency in VTOL and hover modes. This thrust inhibiting effect, called "download", is an important design constraint for tiltrotor aircraft, since it degrades aircraft flight performance during VTOL, hover, and at low forward airspeeds of typically less than 30-40 kts, dependent on rotor/nacelle angle. The swept rotor area that produces the download effect on the wings during VTOL/hover operations can be seen in Figure 1.

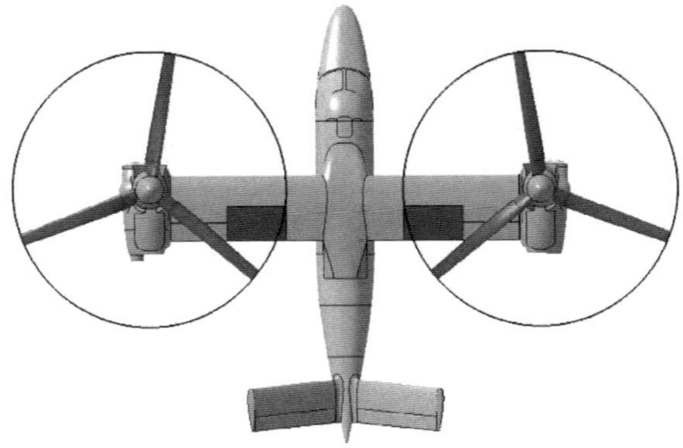

Figure 1: Rotor sweep and effective download area on wings

Historically, tiltrotor aircraft have minimized download with tilting nacelles, full length flaperons and/or separate flaps and ailerons, and limited wing airfoil dimensions (chord length); however, when designing Leonardo Helicopter Division's (LHD) next generation of tiltrotor, non-tilting engines presented significant reliability and operational advantages. Conversely, these advantages were coupled with a significant increase in download, increasing the previous platform's value of approximately 8% to a higher value of approximately 15% on the new platform. To exploit the advantages made available by utilizing the non-tilting engine design, a way to compensate for the increased download was required; a design problem to create a wing morphing system emerged. This system, which can be advantageous to both fast forward flight and VTOL/hover operations, was designed to allow for a chord-wise larger-than-flap surface to articulate, reducing the wing's under rotor footprint in VTOL/hover, while remaining a part of the lifting surface while in horizontal flight. This system has been difficult to implement in the past due to stability, spacing, and structural concerns. Despite these obstacles, a system which implements non-rotating engines has significant enough advantages that the effort to design a complex download reduction system to address the issues created by this change is preferred, when compared to legacy tiltrotor designs.

When looking for new tiltrotor designs, several objectives of the CS2 initiative were found to tie in with future tiltrotor development: reduction in CO2 emissions, reduction in noise, reduced cost of ownership (operations, maintenance, etc.), higher operating efficiency and productivity, and a desire to operate with fast forward speed. Given these objectives, the NGCTR-TD project was created as a research and demonstration platform to showcase advancements in the CS2 objective areas. The goals of the NGCTR-TD project are then to: prove new architectures, technologies, and operations; develop collaborations and partnerships for systems and subsystems, in order to advance the state of the art; and to sow the seeds for future technologies and products. A set of key enabling technologies have been identified, for maturations and demonstration on the NGCTR-TD, under the CS2 program. One of these key enabling technologies is a new wing concept, which has led to the development of the MoWS.

In the Next Generation Civil Tiltrotor – Technology Demonstrator (NGCTR-TD) program, a systems engineering (SE) approach has been used and orchestrated by the SE lead, in the chief project engineer (CPE) office. The SE approach was supported by utilizing requirements management tools, modeling and simulation tools, analysis tools, and a concurrent engineering methodology, applied at both the platform and major systems (structure, transmission, electrical, etc.) levels. Design activities, including tradeoff studies and development of simulation models, were performed by an integrated, multidisciplinary team lead by a focal point (SE lead). Preliminary and critical design reviews (PDR and CDR, respectively) on a system-level were held in order to measure the progress against the platform-level design and customer expectations.

This approach allowed engineers to harmonize and integrate project requirements within a set of constraints, resulting in a cohesive design, through: collection and tracking of customer requirements, pseudo-live simulations of design outcomes during tradeoff analyses, varying operational concerns, reliability concerns, and safety requirements. From the inception of the project, a complete list of project risks and possible mitigations was kept as a living document. Using a holistic approach provided by a SE-driven design methodology, the identification of main stakeholders, formulation of a development/ validation plan, and prototyping/ testing was taken into account throughout the process, using typical systems thinking.

Systems engineering design approach

The morphing wing is identified as a complex system, not only because of the multidisciplinary characteristics and the close interdependencies between the functions/behaviors, but also because of the strong links between the detailed design of low-level physical components (i.e. the up-stops) and the system-level attributes and functions (i.e. stability and controllability). This required a dedicated effort to model the detailed design features during the early phases of the design, developing digital models specifically able to integrate and simulate the impact of low-level component properties onto the MoWS, as well as show the impact on aircraft behaviors.

It is worth noting that creating the digital models enabled the team to "deep-dive" into portions of the detailed design during the initial design phases; something not usually achievable during normal conceptual design phases of a complex design problem. In the early design phases, the system-level and main subsystem-level requirements and functions are established, but the detailed design in terms of low-level components is typically at an early stage. Here, the capability to quickly iterate through the design space is crucial to produce a design solution space that maximizes performance and minimizes redesign later on in the program, as well as reducing program risk and increasing product quality.

However, highly detailed modeling required to give quality feedback on the design's performance is itself a complex task, since the detailed design of components is customarily performed with isolated component analysis once the majority of the design is fixed. Also, the verification of many of the requirements generally only happens later in the development phases and most of the verification is dealt with using experimental test rig(s) well into the design cycle. Often, this is costly and left only as a final verification of components/design on the back side of the systems V. However, this complexity and cost has been reduced by taking advantage of dedicated, multidisciplinary, digital models able to capture many performance outcomes from say the detailed design of a hinge made the structures team, up to the aircraft level loading conditions coming from aeromechanics simulations. Fundamentally, the set of digital models served as a virtual test rig, without the hassle or cost of the experimental part of typical verification tests.

This set of models enabled an agile and highly integrated design that significantly decreases chances of later design fixes, due to the ability to conduct rapid design iterations with concurrent performance feedback from the models. Additionally, the efforts placed into the development of comprehensive simulation models, both at the system and aircraft levels, allowed the team to alleviate the experimental verification and validation test plan, thus avoiding the design and manufacturing of costly and time-consuming test setups.

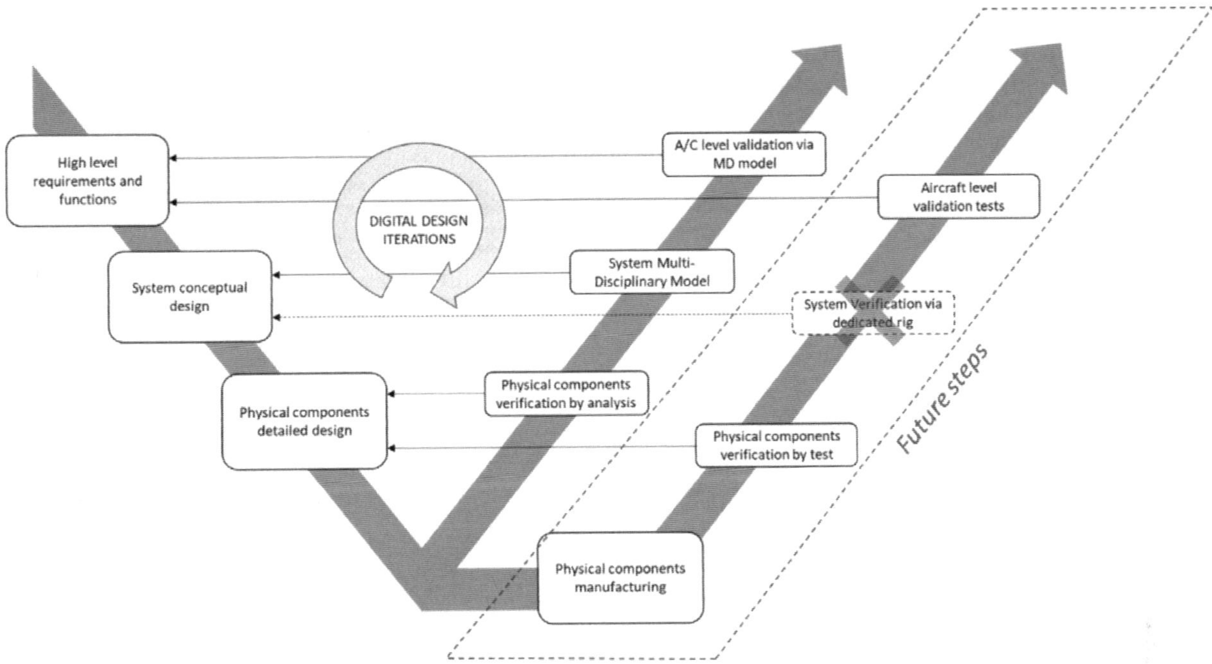

Figure 2: Augmented Systems Engineering "V" with Concurrent Multidisciplinary Modeling Track

Illustrated by Figure 2 above, the team used an augmented systems "V" during the design of the MoWS. Here, the typical right-side of the V, which is generally the manufacturing and verification of components and assemblies, was replaced with a digital version – that which is represented by the detailed, multidisciplinary models. This allowed the team to have pseudo V&V of design in tandem with the early stages of the design process. As this project was focused at only the design of the MoWS, the typical right side of the V has not yet been completed, but is left as future work to build a physical prototype and test in order to compare against the digital twin models.

Throughout this paper, the authors will refer to the entire system as the morphing wing system (MoWS), which is the integration of: a wing surface element, the actuator and its associated mechanical/hydraulic connections, and the flight control system (FCS) to control the movement of the wing surface.

DESIGN PROBLEM DEFINITION

The main objective of the MoWS, as described in the Introduction, was to reduce rotor download on the wing by articulating a wing surface element to a high angle of deflection in flight conditions where rotor wake could impinge on the wing surface. These conditions were identified as VTOL, hover, and at low airspeeds. In order to achieve this, the main objective was broken into the main functions that the MoWS must perform, as is described below.

System functions:

In Table 1 below, the main functions allocated to the MoWS at the beginning of the design activity are summarized:

Functions	F1	Articulate the wing surface
	F2	Carry external loads
	F3	Maintain deployed/stowed position
	F4	Provide stiffness

Table 1: Morphing Wing System main functions

F1: The morphing surface must reduce the wing footprint during VTOL and hover phases by articulating the wing surface element out of the rotor downwash.

F2, F3: The morphing surface is a large aerodynamic surface that represents a non-negligible portion of the wing's span and chord. For this reason, load-carrying and position-keeping capabilities are fundamental functional requirements of the system. The aircraft must be able to sustain aerodynamic loading and maintain a desired aerodynamic wing shape in all conditions (fully deflected, during conversion, and fully stowed).

F4. Aeroelastic stability requirements are applied to the design, in both nominal and failure conditions. The aeroelastic behavior of a movable surface is driven first of all by its mass properties and the stiffness of the connection between the surface itself and the fixed structure (i.e. the wing). Since the morphing surface should be designed with minimum weight and inertia, with respect to the hingeline axis, the connection stiffness is then the primary driver for the design. Thus, for every portion of the flight envelope, the MoWS must guarantee a minimum value for connection stiffness (mainly the rotational stiffness about the hingeline axis and the translational stiffness in the vertical direction), in both nominal and failure conditions.

Functions were then allocated to subsystems, per Figure 3 below. The astute reader will notice that the graphic below includes new functions not identified in Table 1; this is due to the fact that new functions were identified only later on in the design process, once the solution started to take shape. Using iterative design and the systems V allowed the team to then add the new functions to the allocation table during successive design iterations and design the required subsystems needed to perform those system functions, which is shown below.

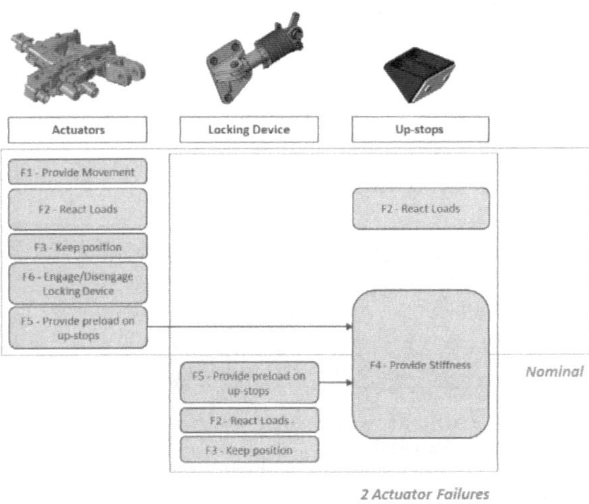

Figure 3: Functional allocation to subsystem components

DESIGN CONSTRAINTS

As with all design problems, engineers must deal with the fact that they are not completely free to design a system without certain constraints being imposed. Below, imposed design constraints, along with their implications on the MoWS design, are described.

Non-tilting engines:

Part of the reason that the investigation into the effect of download on the wings came about was due to an early aircraft-level design decision – fixing the engines in the horizontal axis and only tilting the rotor. Previous tiltrotor designs had the engine and rotor assembly tilt as a single nacelle. This non-tilting engine design presented many operational advantages to the tilting engine design seen in previous generations of tiltrotors (like the AW609 and V-22), though introduced two main drawbacks: 1) loss of residual thrust from the jet exhaust when the engine was tilted upwards, and 2) an increase to the download effect since the entire engine nacelle footprint was now located below the rotor wash (in VTOL mode). Examples of both a tilting and non-tilting engine design for tiltrotor aircraft can be seen in Figure 4

Figure 4: Example of tilting (top) and non-tilting (bottom) engines in tiltrotor aircraft

In previous tiltrotor designs with a fully tilting rotor and engine nacelle, the typical download on the wings was around 8%, which was seen as a good design; note that conventional helicopters usually see approximately 5-6% download from rotor wash on the fuselage and any appendages. The choice of moving to the non-tilting engines immediately resulted in an increase in download to approximately 15%. Given this much higher value, it was evident that download needed to be addressed and was then the basis for this design project. This spurred the need for a solution to the download problem, by reducing the footprint under the rotors, during VTOL and hover operations.

Available space in the wing profile:

The wing box is designed to maximize fuel capacity (high-level customer requirement) and wing torsional stiffness (for aeroelastic stability). These factors meant that the wing box design was already well established and therefore influenced the design of the MoWS such that the actuators are required to be external to the wing profile (i.e.: no parts of the kinematics sit within the wing box). Although external actuators present a source of aircraft drag, a redesign of the wing box design would have been difficult and costly without presenting significant functional advantages.

Actuator selection:

For commonality of parts, reduced inventory/spares needs, and overall system integration, the actuators have been constrained to be the same as those used for other LHD TR platforms. This led to limited design flexibility, but saved time and budget on verification of the actuator itself, since it is a proven part. Reuse of the actuator type. There were several implications on the design of the system. First, it was not possible to freely choose the actuating arm that would move the wing surface, since this was a consequence of the available actuator travel and the deflection angle range required of the morphing wing surface. Also, actuator stall load limits were constrained by the choice of actuator, thus for some parts of the flight envelope, load protection structural elements were needed in order to avoid exceeding actuator load limits.

Further, actuator stiffness is one of drivers for the aeroelastic behavior of the surface and was dictated by current actuator design, so other design elements were required to maintain aeroelastic properties of the original wing. Finally, given the type of actuators and hydraulic system design, the maximum number available for each morphing wing surface was limited to two.

Partial vs full span control surfaces:

As far as mechanical complexity is concerned, allocating distinct physical components to the roll control function and to the rotor downwash release function tends to increase both the design and maintenance costs, mainly because these components are subjected to substantial aerodynamic loads and thus, dedicated fail-safe features are needed to minimize the aircraft level impacts in off-nominal conditions. However, design costs were kept low by utilizing already qualified components; whereas maintenance costs, although a critical aspect for a commercial aircraft, was not considered a key metric for the morphing wing design within the NGCTR-TD. Reducing the download, which effectively increases performance and fuel efficiency in VTOL and hover modes, also has the cost benefit of reducing operational costs by way of fuel savings for every sortie. Over the lifetime of an aircraft, this greatly outweighs the increase in complexity and slight increase in upfront component costs.

Given the complex aerodynamics of rotor wash, a large portion of the download tends to occur in the middle to tip region of the wingspan. Due to this fact, there is then a tradeoff between the size of the flaperon and size of a morphing wing surface, which needs to balance download reduction, roll controllability and performance, wing aeroelastic/structural interfaces, and system complexity. Here, increasing the span of the flaperon may increase performance and controllability, but reduces the ability to reduce the download. Conversely, increasing the size of a morphing section can help reduce download at the cost of performance and controllability.

On one extreme, in the case of a nearly full-span morphing surface (i.e.: a higher priority is placed on the download reduction requirement), it would have resulted in a significantly large chord with a negative impact on the wing aerodynamic efficiency. However, this aspect alone balances the cons related to a dual component approach, even with the loss of performance that would be realized. Therefore, in order to get to a final decision, a third aspect related to controllability was considered in the trade-off. Further, having an outboard aileron solution leads to an aircraft behavior, in terms of roll input vs yaw response, that more closely resembles a conventional airplane. This was considered an opportunity to simplify the development of the fly-by-wire control laws.

More specifically, the final tradeoff considered was between increasing the degree of mechanical complexity and having better download performance by adding the MoWS or finding a compromise between download performance and roll control characteristics with a full length flaperon. As this paper suggests, the choice was to go with the former.

MORPHING WING SYSTEM DESIGN

Given the extensive set of digital models built for this design activity, the majority of design effort was spent in detailed design iterations; whereby, the digital twin was used to not only architect the design, but also to verify the design elements against requirements in parallel through each iteration. This required a pseudo-live representation of the kinematics using an analytical approach, while also taking into consideration attributes of aeroelastic stability and limitations on the actuation system. An example of the optimization of the system kinematics as it progressed throughout the process is seen in Figure 5.

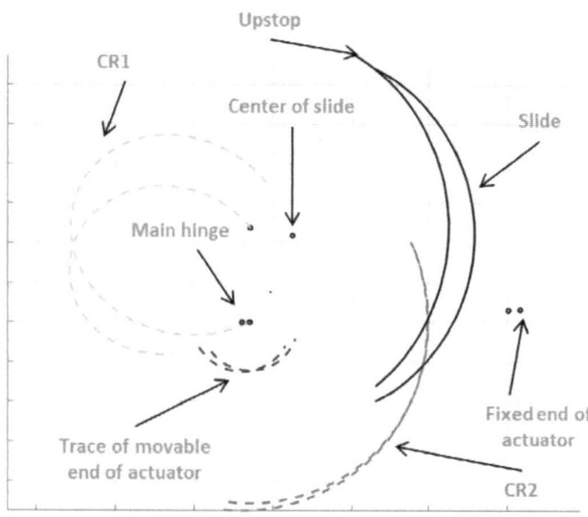

Figure 5: Iteration through the design space during a kinematics optimization process

To accomplish this, a multibody model of the MoWS, comprehensive of actuation system, was developed since the early stage of the design. More than a mere verification tool, this model served almost as a live test bench used to drive the trade-off studies and derivation of some of the lower-level design requirements. The model evolved and progressed into a multibody non-linear model, including the full dynamic-elastic representation of the aircraft – with applied external aerodynamic and inertial loads, and actuation system control dynamics and logic. As can be seen in Figure 6, the models were not only of the MoWS system, but also of the entire aircraft; this was needed in order to accurately calculate and depict the full aircraft forces that would be seen by the MoWS during flight.

Figure 6: MoWS integrated, fully elastic, multibody model

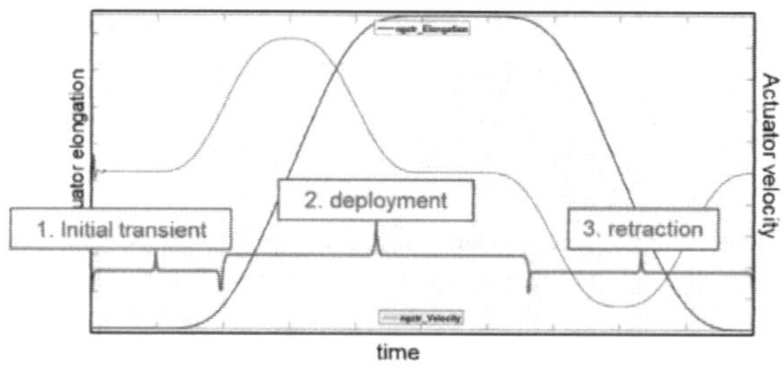

Figure 7: Multibody simulation of actuator travel

Further, non-linear aeroelastic analyses for LCO assessments were used to quantify the requirement on the maximum amount of free play allowable in the system and to evaluate the associated loads for design and structural substantiation. This was substantiated by simulating actuator travel with a digital model, as is seen in Figure 7.

The final MoWS solution is an articulated kinematics scheme, driven by two hydraulic servo-actuators, and controlled by the vehicle FCS. Motion is constrained by a circular track in which two rollers move. Load-carrying limits of the actuators required the introduction of mechanical unilateral up-stops on the wing box (two per MoWS) in order to protect the actuators from trailing-edge-up aerodynamic moments (F2 functional requirement also allocated to up-stops). Forces on the MoWS and up-stops, resulting from the trailing-edge-up and trailing-edge-down aerodynamic forces, are depicted in Figure 8 and Figure 9, respectively.

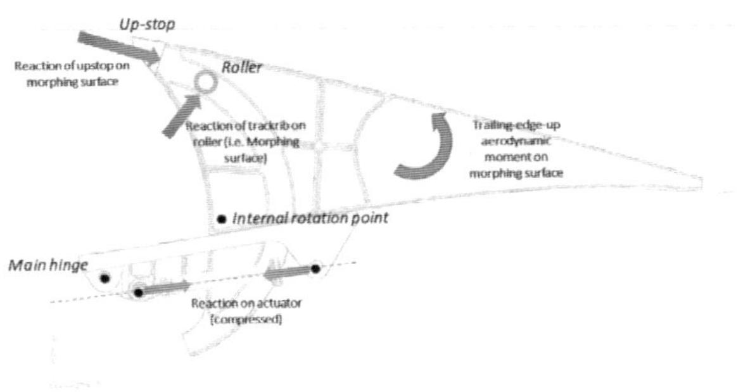

Figure 8: Simplified 2D forces scheme with trailing edge up aerodynamic moment

These additional structural paths on the up-stops provide an increase in required stiffness for aeroelastic stability of the surface at high speeds. With an active control strategy, the contact on the up-stop is secured through a preload applied by the actuators. Here, the goal is to generate sufficient preload on the up-stops in order to guarantee contact under the morphing surface trailing-edge-down aerodynamic moment; this moment acts in the direction of reducing the preload from the up-stops, which could effectively move the wing surface from its stowed state and cause unwanted lift characteristics and aerodynamic instabilities.

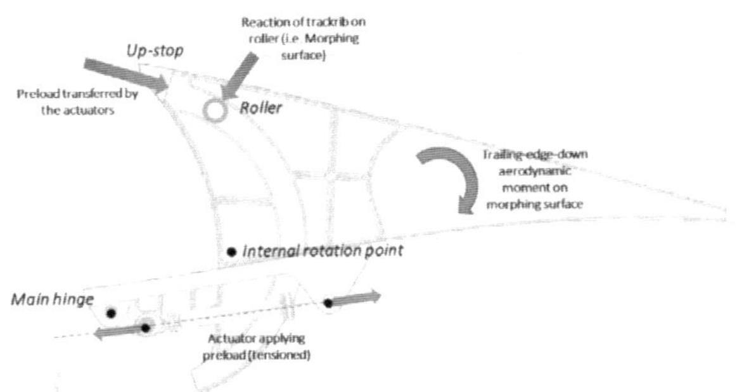

Figure 9: Simplified 2D forces scheme with trailing edge down aerodynamic moment

In a worst-case scenario of a double hydraulic failure, a backup solution is implemented which consists of two passive mechanical locking devices placed close to the up-stops. The locking mechanisms keep the morphing surface in the stowed configuration and additionally apply a preload on the up-stops to secure maximum surface stiffness. The devices are engaged and disengaged by the roller while moving in the last portion of the track when the surface is brought to the stowed position. In a similar fashion, the devices are disengaged in the opposite motion; thus, the actuators are required be able to apply sufficient force to perform this disengage operation. Given that the locking mechanisms are purely passive systems, their preloading capability is a function of an internal spring system, which must be counteracted by the actuators in the disengage operation. This force applied by the passive locking mechanism is then applied as a new constraint on the actuators and was reintroduced into the design problem for future design iterations.

A 3D model of the NGCTR-TD aircraft equipped with the morphing wing design (green) can be seen in Figure 10, below.

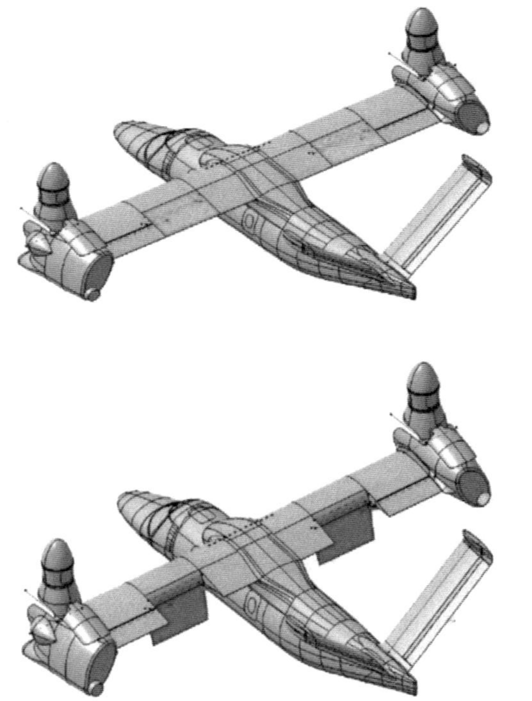

Figure 10: MoWS on the NGCTR-TD, in both the stowed (top) and deployed (bottom) positions

CONCLUSIONS

Utilizing a multidisciplinary design team in conjunction with the pseudo-live simulation models approach resulted in rapid design convergence and the successful design of a complex, articulating, wing surface, which would have otherwise required a longer schedule and at a higher cost. Moreover, the use of a highly integrated set of models for concurrent simulation reduced the need for expensive test rig design, and building/testing of the wing element during design iterations and optimizations[5]. This was due to the fact that using the digital twin models during the design phase allowed at least partial design verification and validation through the design process, rather than solely at the end.

While actual flight testing has not commenced for the design, the predictive models showed that implementation of the morphing wing design on the NGCTR-TD could result in an increase in VTOL and hover performance by at least 5-8%, with respect to power required. It can also be said that the cost savings realized by the digital twin method was on the order of hundreds of thousands of dollars, when considering the test rig, tooling, test bed, analysis, etc. costs of a testing program needed if the pseudo-live simulation was not used during the design process.

Further research in the area of rotor downwash mitigation to reduce download on wings of tiltrotor aircraft is required, especially as industry sees more concepts for future vertical lift (FVL) and urban air mobility/advanced air mobility (UAM)/(AAM) aircraft – many of which utilize a multi-rotor design with tilting propulsion systems. Building on the findings of this paper, further detailed design work is still to be carried out and the simulation model results verified by hardware in the loop and/or wind tunnel testing. Other variations of the design to reduce wing download is also considered as future work and part of the larger CS2 initiative.

AUTHOR CONTACT

Damian Rogers,
Damian.Rogers@LeonardoCompany.us,
Claudio Punzi, claudio.punzi@leonardo.com,
Alberto Angelo Trezzini,
albertoangelo.trezzini@leonardo.com,
Pierre Abdel Nour,
pierre.abdelnour@leonardo.com

ACKNOWLEDGEMENTS

The authors would like to thank the European Union Joint Undertaking (EU JU) for helping to fund the Next Generation Civil Tiltrotor – Technology Demonstrator (NGCTR-TD) project, which spurred the need for this design research. They would also like to thank the TWING Consortium for their generous time and analysis contributions during the design of the morphing wing system.

REFERENCES

1. Acree, C.W. Jr., H. Yeo, and J. Sinsay, "Performance Optimization of the NASA Large Civil Tiltrotor," International Powered Lift Conference, London, UK, July 22-24, 2008.
2. Acree, C. W., Jr., "Integration of Rotor Aerodynamic Optimization with the Conceptual Design of a Large Civil Tiltrotor," American Helicopter Society Aeromechanics Conference, San Francisco, CA, January 20-22, 2010.
3. European Union Aviation Strategy, "Clean Sky 2 Joint Undertaking: DEVELOPMENT PLAN", 28 October, 2021. https://clean-aviation.eu
4. European Union Aviation Strategy, "Technology Evaluator: First Global Assessment 2020", May, 2021. https://clean-aviation.eu
5. Ulmer, T. and Amin, J., "Virtual Testing of High Lift Systems," SAE Technical Paper 2013-01-2280, 2013. https://doi.org/10.4271/2013-01-2280

Simulation and Integration Laboratory (SaIL) Testing of Integrated Cockpit and Fly By Wire

Luca Belluomini
FCS Analysis & Integration Lead
Tiltrotor Technologies
Leonardo Helicopters
Philadelphia, PA, USA

Sid Xiao
AW609 Handling Qualities Specialist &
Technical Leader
Leonardo Helicopters
Philadelphia, PA, USA

ABSTRACT

Modern vertical lift aircraft are complex and highly integrated systems that require careful design to guarantee safety and reliability. The Leonardo Helicopters AW609 Tiltrotor is one of the most complex and integrated systems in civil aviation. The AW609 relies on a full-authority, high-bandwidth, triplex-dual redundant architecture, fly-by-wire system which enables the flight crew to control transitions from helicopter to airplane mode, control the flight path of the aircraft, and manage the engine power demand. The AW609 Vehicle Management System (VMS) can manage and mitigate, in real-time, the system failures to ensure continuous safety of flight. The AW609 VMS is designed to a stringent set of requirements and objectives that meet or exceed the civil certification rules and allow the aircraft to satisfy its marketing objectives. As a result, many complex and highly integrated functions are integrated into the VMS and require rigorous methods of design assurance. This paper describes the use of hardware-in-the-loop simulation to verify qualitative safety assessments for Flight Control System (FCS) failure conditions that would be unsafe or impractical to fully evaluate in flight. Using the Simulation and Integration Laboratory (SaIL), it is possible to examine aircraft transients caused by a failure, crew recognition of the abnormal event, recoverability after the failure transient, and continued safe flight to landing in a thorough and efficient manner without presenting a safety risk to the flight test aircraft. The data illustrate that SaIL is an effective means to identify system problems and define emergency procedures that might be otherwise undetected and undefined from bench testing and stand-alone simulation. The net result of implementing SaIL methodology is an increase in level of system complexity and functionality without a corresponding increase in flight test time, safety risk, and cost.

NOTATION

APSW	Application Software
CAS	Crew-Alerting System
CCDL	Cross-Channel Data Link
EHV	Electrohydraulic Servo Valve
EMD	Engineering and Manufacturing Development
FAA	Federal Aviation Administration
FCC	Flight Control Computers
FCS	Flight Control System
FCU	Fuel Control Unit
FHA	Functional Hazard Assessment
GTR	Generic Tilt Rotor
HMI	Human Machine Interface
ICD	Interface Control Document
IFR	Instrument Flight Rules
IMC	Instrument Meteorological Conditions
LRU	Line-Replaceable Unit
LVDT	Linear Variable Differential Transformer
NFE	Normal Flight Envelope
OAT	Outside Air Temperature
OFE	Operational Flight Envelope
SaIL	Simulation and Integration Laboratory
TLTI	Triple Lab Tie-In
VDT	Variable Differential Transformers
VFR	Visual Flight Rules
VMS	Vehicle Management System
VTOL	Vertical Take-Off and Landing
WOW	Weight On Wheel

Presented at the VFS Development, Qualification and Affordability of Complex Systems, Huntsville, Alabama, USA Sept. 14-15, 2022. Copyright © 2022 by the Vertical Flight Society. All rights reserved.

INTRODUCTION

The AW609 (Figure 1) answers the need for an aircraft with the speed, range and altitude of a fixed-wing turboprop airplane and the vertical take-off and landing versatility of a helicopter. This revolutionary combination, along with the comfort of a pressurized cabin in which is possible to fly above bad weather, makes the AW609 an unrivalled asset.

The Certification Basis for each FAA Type Certification Program constitutes the legal regulations that provide for the minimum level of acceptable safety. The AgustaWestland Philadelphia Corporation's (previously known as AgustaWestland Tilt Rotor Company) AW609 is a tiltrotor aircraft that does not fit into an existing category of air vehicle, and as such, qualifies under the definition of a special class of aircraft per FAA Part 21.17(b) for the definition of certification basis.

The AW609 is designed for flight into known icing conditions and meets the highest FAA transport category requirements for both airplanes and helicopters including single engine operation and autorotation.

The AW609 has a forward swept wing with dihedral attached to the top of the fuselage and the empennage is configured as a T-tail. Wing tip mounted nacelles, each containing engine, gearboxes and a three bladed rotor, tilt from horizontal position (airplane mode) to the vertical position (VTOL mode).

A cross shaft connects the drive system between the two nacelles for sharing power and synchronizing proprotors speed. Proprotors provide pitch, roll, and yaw control in VTOL / conversion modes and yaw control in airplane mode. Wing flaperons provide roll control (ailerons) in airplane mode and downwash reduction in VTOL mode. A tricycle retractable landing gear consists of a castering nose dual-wheel gear and braked main wheels. The nose and main landing gear retract hydraulically into non-pressurized bays.

The power plant consists of two Pratt and Whitney PT6C-67/A engines designed for operation at all nacelle positions. The engine fuel control system is hydro-mechanical with electronic engine limiting and automatic startup functions provided by the three aircraft Flight Control Computers. Fuel tanks in wing consist of ten crash resistant bladders installed in cells between wing ribs. Fuel is fed to the engine by an airframe mounted boost pump system, while transfer between tanks is achieved by gravity and jet pumps. Fuel is transferred between the wings by a cross transfer pump.

Figure 1 - AW609 Leonardo Helicopters

The AW609 rely in its vehicle management system (VMS) which consist in a digital VFR/IFR avionics with dual pilot capability and a fully digital triple redundant fly-by-wire flight and engine control system (FCS). The AW609 VMS reduces pilot workload, maximizes situational awareness, and improves the flight safety and fault tolerance.

The Integrated Avionics System is built around Rockwell Collins Pro Line Fusion. This system incorporates:

- Three Adaptive Flight Displays (One Primary Flight Displays for the pilot and the co-pilot each, one for Engine Indication and Crew Alerting System located centrally in the instrument panel)
- 2 Cursor Control Devices (Multifunction Pilot and Copilot)
- One Multifunction Keyboard

Over the past two decades, tens of thousands of hours have been spent testing the AW609 Vehicle Management System (VMS) equipment. Individual components and software items have been tested and qualified at vendors and in stand-alone laboratories. The overall system has been integrated and tested in the Leonardo Helicopters Simulation and Integration Laboratories (SaIL), during flight tests on the first 5 AW609 Engineering and Manufacturing Development (EMD) aircraft, and during engine controls integration in the PT6C-67/A engine test cell at Pratt & Whitney-Canada.

Flight control laws and redundancy management algorithms have been extensively evaluated using piloted simulation and mathematical models.

Previous aircraft development programs have cited the benefits of SaIL (a.k.a. "iron bird" or "hot bench") testing (Refs. [1], [2], [3]). This paper presents the AW609 VMS as a case study for the use of SaIL as an asset for developing a complex, highly integrated flight system. First, the paper presents an overview of the AW609 VMS and its integrated functions. The paper then describes the planning and execution of SaIL with a summary of the significant results. The paper concludes with a list of lessons learned from a comprehensive SaIL test program on the AW609 VMS. These findings highlight various methods to maximize the benefits of SAIL testing.

VMS DESCRIPTION

The vehicle management system (VMS) is defined as the integrated aircraft system that provides functions necessary to control and manage the aircraft. The VMS encompasses aircraft system elements that transfer information between the flight crew and the aircraft, necessitating integration as a single system.

The VMS enables the AW609 flight crew to control the transition from helicopter to airplane mode, control the flight path of the aircraft and manage the engine power demand the aircraft, using conventional controls. The conventional controls consist of center cyclic stick, pedals, power lever (which is configured with the same position and functionality as a collective stick) and nacelle controls. This approach seeks to minimize the additional familiarization and training required for flight crews transitioning from airplanes and helicopters to tiltrotors. This approach has been proven through flight experience with the conventionally configured cockpit of the XV–15 tiltrotor demonstrator whose guest pilots from a variety of flying backgrounds have adapted quickly to flying a tiltrotor (Refs. [5], [6], [7]).

Additionally, the VMS provides the integrated function to manage the aircraft's flight critical, mission critical, and mission support subsystems, to annunciate flight, mission, and system information to the flight crew, and to provide data collection and monitoring of aircraft parameters for use by maintenance and operations personnel.

The VMS architecture is driven primarily by flight safety requirements. In order to meet SAE ARP4761 [8] flight safety requirements, the Flight Control System FCS (Figure 2) is composed of three identical and redundant flight control computers which communicate over the Cross-Channel Data Link (CCDL). Each flight control computer has dissimilar processing lanes running parallel computations of the Operational Flight Program for additional design assurance and fault tolerance.

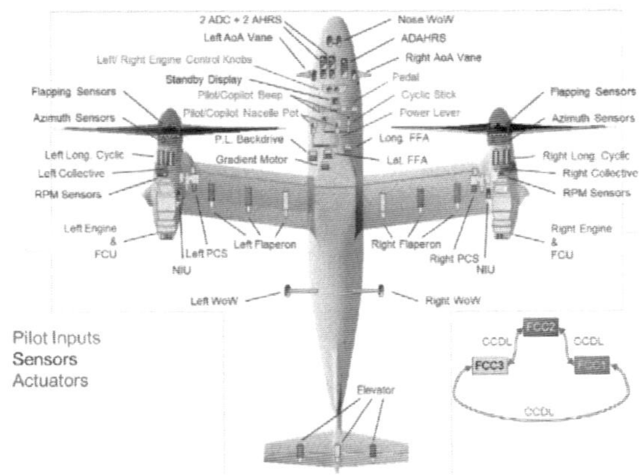

Figure 2 - Top level configuration of the AW609 Flight Controls System

Three identical FCCs (Flight Control Computers) continually perform extensive inline monitoring (in hardware and software) and where necessary, compare their local data against the data from the redundant channels to detect random hardware and interface signal failures. Where practical, the FCCs use majority voting or signal averaging to maintain consistency between the separate channels. Within each FCC, there are two computational lanes, a primary lane, and a secondary lane, configured in an active/standby mode of operation. A single FCC, operating with the primary lane or secondary lane active, is capable of safely landing the aircraft although some non-critical functions may be lost in this degraded mode of operation.

The FCCs provide interfaces and control for FCS sensors, actuators, and data busses. They provide the digital processing for signal selection, Control Laws, actuator control and monitoring functions.

Pilot inputs are sensed by triple redundant Variable Differential Transformers (VDTs) connected to the cockpit controls for each axis. Triple redundant VDTs are similarly provided to sense flap and power lever position.

The nacelle control command is sensed by Switches (with resistors for inline monitoring) at each pilot station or the overhead backup switch. Several discrete switches and panels are also provided in the cockpit to allow the pilots to select the appropriate operating states for the FCS.

Force feel for the pilots is provided by arrangements of springs and dampers. For each of the pitch or roll axes one spring is provided connected through a magnetic clutch in an arrangement that allows a variable force gradient by varying the mechanical advantage against the spring using a gradient change motor. For the yaw axis, a single spring gradient is provided. In the collective axis, a pilot adjustable variable friction device is provided. In the collective, pitch and roll axes, trim actuators are provided.

Conventional air data sensors, rate and attitude sensors, RPM/Azimuth sensors, flapping sensors, sense aircraft states. Multiple redundant transducers also sense control surface positions and nacelle actuator positions. Discrete sensors are used to sense certain aircraft configuration changes that are necessary to appropriately configure the operating parameters of the FCS, such as Weight on Wheel (WOW) sensors and hydraulic system pressure switches.

The outputs of these sensors are converted to digital format by redundant FCCs, which then compute the desired control surface commands, rotor commands, nacelle commands, and engine power demand. These commands are then converted to analog format to be used as commands to the electro-hydraulic surface, rotor, nacelle, and engine actuators.

Redundant actuators, with three actuators per surface, control each of the control surfaces. The rotor control actuators for collective pitch are also triplicated, but the outputs are mechanically force summed as part of the actuator assembly, and only a single mechanical connection is made to the collective spider and swashplate, and at the fixed end to the transmission. Similarly, the rotor control actuators for longitudinal cyclic pitch are triplicated, with the outputs mechanically force summed for a single mechanical connection to the swashplate. Lateral cyclic pitch is fixed. All the actuators are close loop controlled from the FCCs, using actuator position feedback.

Collective and Longitudinal swashplate actuator assemblies are triplex; hence, each assembly is composed by three actuators tied together on a single output flange; each actuator in the assembly is commanded by one of the three FCCs and powered by a single hydraulic line.

Flaperon and Elevator surfaces actuation is provided by three separated simplex actuators for each surface; each actuator is commanded by one of the three FCCs and powered by a single hydraulic line.

The nacelles are normally controlled by a pair of hydraulic drive modules that drive into ballnuts which drive the telescoping ballscrews which move the nacelles. The hydraulic drive modules are connected by a cross shaft which provides for nacelle position synchronization. The modules include a primary hydraulic motor drive and a similar backup hydraulic motor drive.

Engine power demands are provided from the FCCs to hydro-mechanical Fuel Control Units (FCU) on each engine. These power demands take the form of analog voltages that drive torque motors on the FCUs through closed loop servo drives. These torque motors comprise the first stage of two-stage EHVs that in turn control servos that position the engine throttle input. The FCC servo drive position control is provided through LVDT sensors, which feedback FCU servo position to the computers.

As such, the FCC inputs are the electrical equivalent of the mechanical throttle inputs used in traditional engine control systems.

With all three FCCs providing inputs, the torque motor normally operates on the sum of the magnetic flux generated by the three servo-drives. When one of the three FCCs servo fails, the command portion that was supposed to be provided to the torque motor will be supplied by the other two FCCs, to keep the throttle response constant. There is a small transient, due to the time required to detect the failure and to compensate the command. The servo operates using a source of regulated fuel pressure that is provided from the FCU. The servo range is mechanically limited in such a way that the throttle position cannot command the engine below idle power or above the gas generator speed redline. The torque motor is provided with a mechanical bias that is usually overcome by the current command from the FCCs. This bias causes the throttle position to gradually retard to the idle position if control of the torque motor is lost. This function provides a repeatable and predictable response to torque motor command failures or any other failures that would cause the FCCs to shut down their commands to the FCU.

In addition to providing control surface position commands, the FCCs will also provide the capability to monitor and manage the individual system components. In the case of individual failures, the FCCs will disable the affected components to maintain safe and effective control of the aircraft using the system redundancy. In the case where such actions are necessary, the FCCs will also alert the pilots to the failures and provide indications to identify the necessary maintenance actions.

Requirements	Speed Stability
	Lateral–Directional Static Stability
	Dynamic Stability
	Handling Qualities
	AeroServoElastic Structural Mode Filtering
Features	Pitch, Roll, Yaw Rate Stabilization
	Authority Limited Pitch Attitude, Roll Attitude and Heading Stabilization
	Model following Rate Command
	Lateral acceleration feedback and turn coordination
	Set and Forget Torque Command
	Airspeed envelope cueing
Unique AW609 features	Full Authority, Full Time Stabilization
	Phasing and Mixing with Nacelle Angle and Airspeed
	Cross–axes branches for response decoupling
	Roll Rate to Total Pedal in Airplane Mode
	Total Lateral Stick to Diff. Coll. Pitch in Airplane Mode
	Engine control and limiting integrated in Control Laws
	Tactile cueing of power ratings and limits
	Conversion protection
	Control Power Management System
	Active Cyclic Load Alleviation in Airplane Mode

Table 1 - AW609 Control Laws Requirements and Features

The functions performed by the Control Laws (Table 1) allow the pilot to control the aircraft safely and efficiently both on the ground and in the air from engine start through shut down. In general, functions performed by the Control Laws are those which are highly integrated with the aircraft system, depend upon aircraft response in a closed loop sense, and require aircraft simulation or flight test and pilot handling qualities assessment to specify their design and demonstrate compliance with their performance requirements.

The Control Laws are designed to operate in a minimum number of distinct modes as a function of pilot selection and system health. Minimizing the number of modes of operation of the Control Laws simplifies the task of integrating and testing the flight control system and simplifies the pilot's task of safely operating the system.

The Control Laws have two main modes (Figure 3), NORMAL (FCS Augmentation "ON") mode and DIRECT (FCS Augmentation "OFF") mode. There are several sub-modes which may be individually engaged or disengaged independent of NORMAL and DIRECT modes. There are also several sub-modes which may be individually engaged or disengaged only while in NORMAL mode. Which mode or sub-mode the control laws are in is determined by a combination of detected system failures and pilot selection.

The Control Laws operating in the NORMAL mode shall provide a system which gives "Satisfactory" handling qualities throughout the Normal Flight Envelope.

The Control Laws operating in the DIRECT mode provide a system which is adequate for continued safe flight and landing. This mode is entered only following a failure which prevents operation in the NORMAL mode. Transition from NORMAL to DIRECT mode may occur automatically as a function of detected system failures or manually by the pilot in the event of an undetected failure by switching the "FCS AUG" switch to the OFF position.

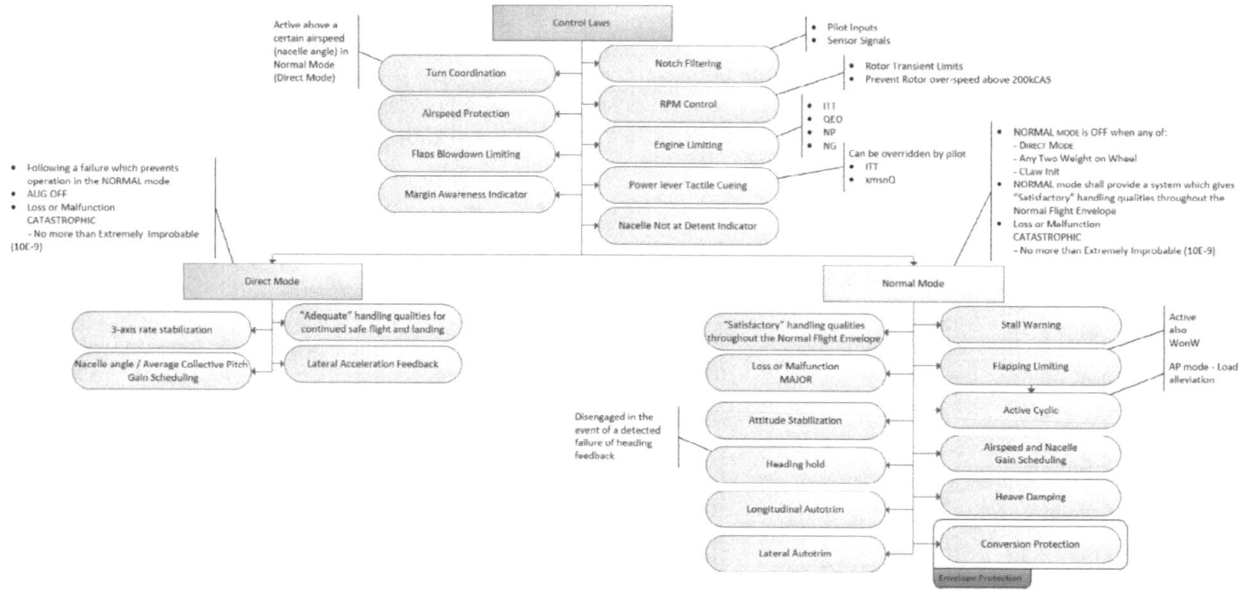

Figure 3 - Control Laws Modes

SaIL DESCRIPTION

The Leonardo Helicopters Simulation and Integration Laboratories (SaIL) provides unique capabilities in terms of software and hardware integration, testing, requirements verification and validation and simulation activities. This lab contains a real implementation of all the Electrical, Avionics and Flight Control System of the aircraft, including the hydraulic system. This means that all the relevant real aircraft equipment for the above systems are interfaced and operated as in the real aircraft, within a representative environment and external conditions that can be simulated as required by the task or function to be tested.

The SaIL is used for highly integrated system testing. The SAIL allows the system to be exercised in ways that are either impractical on the aircraft, or that may cause an unnecessary hazard to the flight test aircraft and its crew, as in the case of critical failure testing.

The highest level of lab integration involves multiple bench tie-ins, using the FCS, Electrical, Avionics benches, Hydraulic RIG and Flight Simulation environment. These tie-ins provide the highest fidelity test capability. The integration is obtained incrementally: a minimum set of components are interfaced and tested to verify operation of a single function, then more components are included and tested until all the flight control components are fully integrated. At this stage, system verification testing may begin. The architecture of the AW609 Simulation and Integration Laboratories is based on a modular and reconfigurable design that allows different levels of testing on the various equipment, subsystems, and systems including standalone, interconnected, or multitasking methods.

The final purpose of the overall structure is to perform Integrated Hardware/Pilot in The Loop with all the real HW represented in the Laboratories that are connected as in the real aircraft, to demonstrate proper functionalities and to support safety of flight clearance on the tested configuration.

The different computing elements share data in real-time over the same data map for an efficient data monitoring and testing performance. The AW609 Simulation and Integration Laboratories is composed by the following five main areas (Figure 4):

1. The Avionics Laboratory composed by Avionics Test System, representative cockpit mockup, visual scenario based on image generation and projection, the necessary test station to perform testing/evaluation on these systems;

2. the Flight Control System Laboratory composed by FCCs Test Systems, the hydraulic rig to install the real actuators in aircraft representative

manner, the hydraulic power generation of the three real hydraulic systems and also for the simulated loads over the actuators, the rotating rig necessary to support changes to the azimuth/flapping system, the necessary test station to perform testing/evaluation on these systems;

3. the Electrical System Laboratory composed by the electrical power Generator Drive rigs, the Power distribution benches, the necessary test station to perform testing/evaluation on these systems, AC & DC electrical load bank facility;

4. the Ice Protection Control Test System composed by the Ice Protection Controller and Distributor Bench, the Power distribution benches, the necessary test station to perform testing/evaluation on these systems, representative ice protection electrical load bank facility;

5. the Simulation for Out-the-Window Visualization composed by a 120-degree field-of-view dome projection system to support advanced visualization, a real-time connection with the system under test.

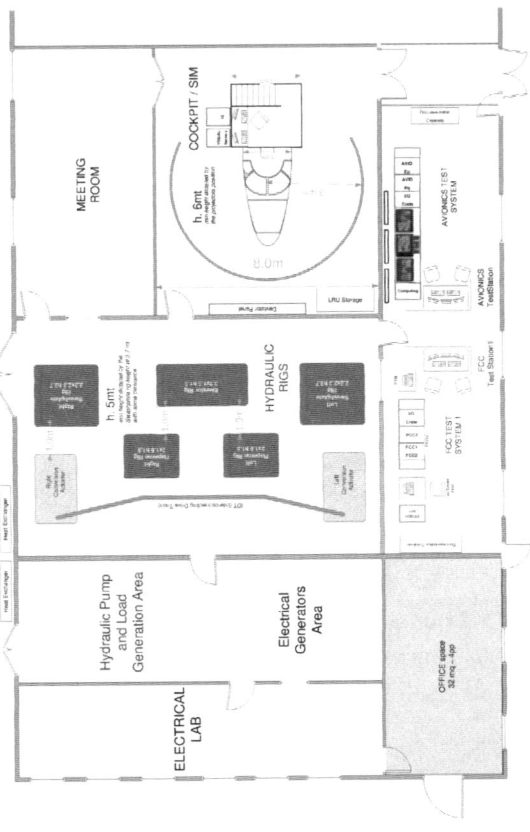

Figure 4 - SaIL Layout (FCS laboratory in yellow)

All the aircraft interfaces and flight related systems are represented in the SaIL, either as real LRU or as aircraft representative models, depending on a cost-effective analysis performed for each of them. The SaIL is where the real FCC's targets are used to run the flight control system software and evaluate its behavior in the integrated VMS. It is available in three different configurations:

1. Stand-Alone Operation Environment: each component of the SaIL can be run and tested independently with inputs being provided by emulated hardware. Examples of the external simulated equipment are hydraulic & electric actuators, cockpit sticks, panels & displays, Inertial Reference Systems (IRS), Air Data, Attitude and Heading Reference System Unit (ADAHRS), Nacelle Interface Units (NIU's), etc.

2. Mixed Real / Simulated Component Environment: SaIL provides connections where the aircraft equipment simulations can be replaced with actual aircraft hardware during testing.

3. Triple Lab Tie-In (TLTI) Environment: this environment utilizes all three aspects the SaIL, the system test benches, aircraft-representative hardware (including cockpit rig), and a projection screen that provides visual feedback of aircraft response as determined by a high-fidelity flight simulation.

Piloted simulations are used in the development of the control laws for the flight control system, including the development and evaluation of mode switching logic. These simulations include the development of envelope protection algorithms and structural protection algorithms, as well as evaluations of their effects on aircraft handling qualities and piloting techniques.

During the system requirements development phase of the program and spiral evolution, piloted simulations and developmental flight testing are used to support determination of functional hazard criticalities for the system safety analysis activities, as well as determination of critical flight conditions for system failure effects and controllability of the aircraft following critical failure combinations.

During the system design verification phase, the simulator is integrated with the real aircraft hardware and software in the system integration facilities to provide a "hardware-in-the-loop" simulation capability. This capability is used to verify the "as built" and fully integrated system operation and the effects of failures. The piloted simulation provides great flexibility and efficiency for testing the large number of failure combinations that are possible.

With the approval of the certification authorities, this capability is also used for demonstrating compliance with the applicable regulations that define the acceptable effects of failures on the flight control system. This requires that, for the conditions to be demonstrated, the SaIL must be shown to be valid for its use as a reasonable simulation of the AW609 type design.

Using the aircraft simulator and the SaIL, information can be gathered on aircraft transients caused by a failure, crew recognition of the abnormal event, recoverability after the failure transient, and on continued safe flight to landing after recovery without presenting a safety risk to the flight test aircraft.

Airframe, drivetrain, and proprotor system dynamics are integrated into the SaIL environment via the SimRX mathematical model. This model was developed initially by Bell Flight and adjusted by AWPC based on flight and wind tunnel data correlation and is based on the NASA Generic Tilt Rotor (GTR) simulation tool used for the development of both the XV-15 technology demonstrator as well as the V-22 Osprey platform.

The model contains mathematical representations of the airframe (separated into Fuselage, Wing-pylon, and Empennage components), rotor, and drivetrain. Effects such as rotor-airframe interaction wash effects, airframe wake effects, and ground effects, are included. The powerplant is incorporated as an external vendor-supplied model.

Airframe Aerodynamic data is derived from scaled wind tunnel data: initial campaign was conducted at Texas A&M University in 1996, a second campaign was conducted at the NASA AMES high speed wind tunnel in 2017.

Proprotor aerodynamic data is derived from legacy tiltrotor data and from various powered rotor wind tunnel campaigns.

The data is supplemented by various studies conducted via CFD and other analytical tools throughout the history of the AW609 development program. Complex wake and wash interaction effects were initially based on legacy tiltrotors and were improved as the AW609 program matured.

The SimRX mathematical model shares a symbiotic relationship with the development of the AW609 design. During development phase, SimRX was used for the initial design the Flight Control System, specifically the Flight Control Laws (CLAWs). Once the EMD aircraft was available, data collected from flight testing was used to improve SimRX model fidelity, while CLAWs were refined using the combination of flight test data and further simulation using the improved SimRX model. In this way both the Flight Control System and the SimRX mathematical tool gained fidelity throughout each iteration of the AW609 program, resulting in maturity for the Flight Control System, as well as high-fidelity for the SimRX mathematical model.

The resulting model can then be used for the requirements definition and system design verification phases with high confidence.

TESTING ACTIVITY

Failure evaluation framework

A Functional Hazard Assessment (FHA) has been performed for AW609 FCS (Figure 5) under guidance from SAE ARP4761 [8].

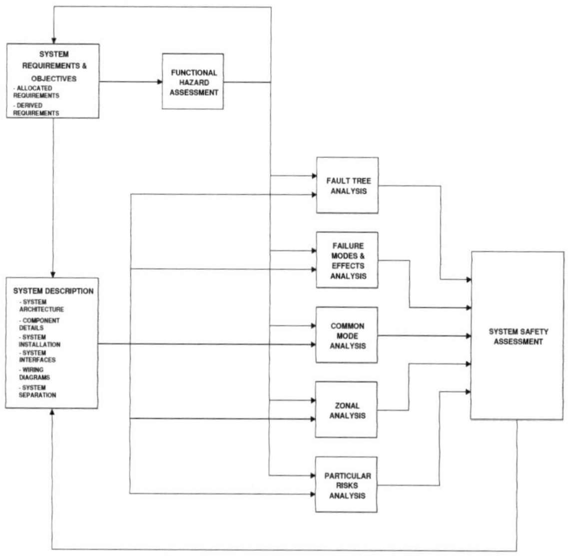

Figure 5 - System Safety Analysis Process

The FHA provides a structured methodology for evaluating system functional failures and assessing their impact on aircraft safety. The FHA is the entry point for the total system safety assessment process required to show compliance with the aircraft certification basis. The FHA process outputs safety requirements which are allocated into the system's hardware and software design. The FHA establishes the type and level of analyses required to substantiate that the system provides the levels of safety necessary for its intended function(s). These subsequent analyses are conducted against the detailed implementation of the system to assure that the system, as built, complies with its safety requirements and with the applicable regulations.

AC#4 First Flight de-risk framework

SaIL Triple Lab Tie-In (TLTI) configuration is used to substantiate the airworthiness of integrated aircraft design and any substantial design changes. First flight of prototype aircraft #4, the first with production-representative avionics configuration, was de-risked by a comprehensive SaIL test program which investigated aircraft failures using pilot and hardware in the loop simulation to achieve the following objectives:

- Verify the FHA classification
- Assessment of Failure transients and recoverability
- Assessment of System reaction
- Verify the human machine interface (HMI) provides appropriate workload mitigation with situational awareness
- Collect data to develop or assess emergency procedures
- Validate the warning cautions and advisories are effective in realistic situations

This testing activity relied on the validated SaIL environment. Functional characteristics of each system element in the SaIL when operating in a closed loop have been verified and integration testing have been performed to verify the interface control document (ICD), the power functionality, and Avionic, FCS system, and mathematical models' interaction. The SaIL flight characteristic has been validated by correlation with the AW609 flight test data.

Functional Failure selection

VMS Functional Hazard Assessment (Figure 6) identifies 13 system-level functions of the Flight Control System – functions such as providing pitch, roll, yaw, and thrust control to the aircraft. For each of the 13 functions, various functional failures are identified which would lead to erroneous operation or loss of that function - more than 200 possible functional failures were identified for the 13 FCS functions.

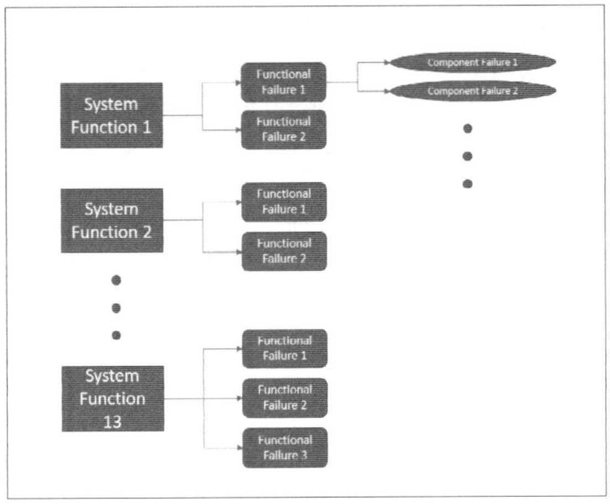

Figure 6 - Functional Hazard Assessment breakdown

For each of the functional failures, a System Safety Assessment has been performed on the VMS architecture to identify the component failures and failure combinations that lead to functional failures identified in the FHA. Since the complexity of some FHA cases involving the combination of multiple functional failures and confronted with the high degree of integration within the FCC, it may be difficult to identify common failure modes by relying only on design review considerations. Therefore, fault trees are built with specific criteria to identify the common failure modes. In accordance with ARP 4761, the fault tree minimal cut has been used to identify the smallest set of primary events which must all occur for the undesired top-level functional failure to occur.

For the specific case of the AW609 AC#4 first flight clearance, VMS system hardware and software changes specific to AC#4 were identified - for example, the Inertial Sensing Units of the previous EMD aircraft were replaced with integrated ADAHRS systems on AC#4 to reflect production design, therefore any failure pertaining to sensor data were selected to be evaluated. Conversely, control actuators were not changed for AC#4, therefore failures pertaining to control actuators were not selected. This process is repeated for all hardware and software changes. In total, 60 functional failures from the FHA were selected to be evaluated.

The effects of each failure are evaluated to ensure the systems and flight crew responded in a manner consistent with the safety requirements.

Flight Condition Selections

Once failures are identified for test, the FHA is used determine the Failure Conditions to be evaluated. The

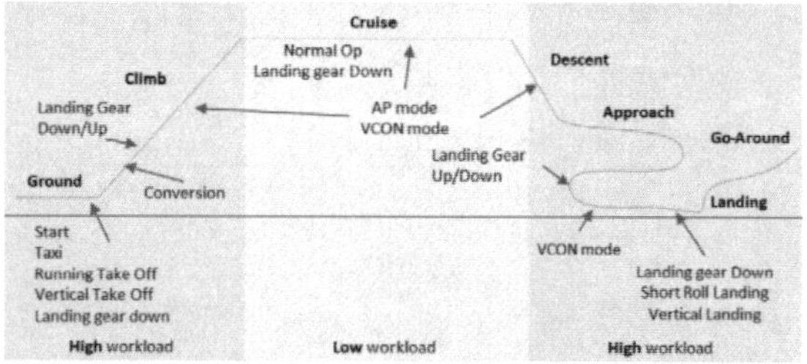

Figure 6 - Operational Phases and related crew workload

FHA identifies the worst-case scenario in terms of Operational Phases (Figure 6), Operational Conditions, Aircraft Configuration, and Environmental Conditions.

This information is used to define the setup (failure scenario) of the test. For example, a failure hazard classification may be worse in VTOL mode in IMC conditions. This may drive the test to have the failure injected during an IFR approach or landing. If the FHA indicates the failure condition is applicable to ALL phases and configurations, then engineering judgment is used to identify when to inject the failure. The AW609 requires two pilots to operate, however testing was conducted with one pilot. Most of the time this should increase pilot workload and make test results more conservative. If crew coordination is deemed via engineering judgement to contribute to making the scenario worse, the test will be repeated with two pilots.

The probability of the failure condition occurring (based on the fault-tree analysis) is considered: any failure condition with an extremely improbable likelihood of occurrence (less than 1e-9 per 100 flight hours) will not be evaluated.

Additionally, considering the specific activity limited to substantiate the integration of the design changes on AC#4, the flight conditions in which the failures are tested were defined based on pilot and engineer judgment and modified (as needed) during the testing phase at pilot discretion. Flight data and physics-based trend considerations (control margins, stick-free damping, control power and bandwidth), additional combinations of GW/CG/altitude/OAT/speed are considered to create the conservative worst-case testing condition for each specific failure being tested.

Under the above-mentioned considerations, over 100 test cases are tailored on the AW609 flight envelope. Each of these tests covers a series of system and equipment failure combinations.

Triplex redundant digital systems result in a large matrix of failure evaluation. These evaluations can be dangerous or even impossible to test on the real aircraft in-flight. Using SalL, the matrix of failure conditions can be systematically evaluated to ensure that the effects are consistent with design requirements.

Failure mode evaluations were conducted in a 6-month period in the SalL environment. The tests covered the full range of single and multiple failure combinations from the following categories: erroneous actuator operation, loss of actuator power, hydraulic failures, power supply failures, processor faults, common mode processor failures, wire and bus faults, cockpit control force/trim malfunctions, display faults, sensor failures, plugged air data ports, sensor hardovers, slowovers, and oscillatory failures, and propulsion system malfunctions.

Test methodology

Selected failures are injected during target flight phases. The pilot reacts to the failure and regains/maintains control of the aircraft, this phase is recognized as Transient Evaluation. Once the pilot has stabilized the flight path of the aircraft following a failure injection, the crew carry out the appropriate Emergency Procedure, should one be applicable. Engineering and pilots determine if the existing Emergency Procedure is correct and appropriate or if modification or addition is required. Recommendations are collected in the dedicated test report and final emergency procedure recorded in the aircraft flight manual. Specific tests are done to investigate failure conditions based on what the initial planned testing revealed.

Several maneuvers and mission related tasks are flown to identify deficiencies in flight characteristics with the failure present. These tasks enable the pilot to evaluate the ability to perform safe flight to landing. They evaluate pilot workload and handling qualities for closed-loop tasks that derive from typical mission tasks as the pilot flies the mission profiles from Figure 6 such as –

- Attitude capture
- Vertical speed capture
- Attitude and heading maintenance
- Low speed helicopter mode translational repositioning

- ILS approaches
- Nacelle conversion/reconversions

Failure Injection capability

The capability of FCS software to modify gains, inject stimuli, or simulate failures is provided both from Control Laws (CLAWS) hard coded injection points and from FCS bench tools and functions. The definition in software of tunable gains and injection points, represent the only points of access to the CLAWS, while in flight using flight test functions. However, in the lab environment using the FCS bench, test scripts, software memory addresses access, actuators and sensors models and wiring patch panels, allows access to any part of FCS Application Software (APSW) and Flight Control Computers (FCCs) interfaced hardware. Failures will also be initiated with the use of the dedicated test bench computer, interface patch panel, and models to set attitudes or sensor readings to unusual values, modify FCC I/O from/to sensors and actuators.

The avionics bench enables the stimulation of failures. Avionics bench tools and patch panel can be used. Some failures can be caused by powering down parts of systems.

The lab is not configured to allow actual hydraulic system leaks however the AW609 does not have leak isolation capability. The impact of a hydraulic system leak can be simulated by shutting down hydraulic pumps.

Mechanical failures are limited to disconnecting some equipment. This cannot be done while a test is in progress and therefore an assessment of transients associated with these failures are not possible. However, evaluation of flight characteristics and workload with the failure present is still possible. An example of such a failure is the disconnection of a flight control viscous damper.

Transient Event Definition

A transient is a change in aircraft motion because of the failure. The FCS system is designed with failure detection time requirements and limits to minimize transients. Requirement verification can be done accurately with requirements-based test methods, out of the scope of this analysis. This testing is to validate the timing requirements, *Response Time* (Figure 8), are correct and prevent loss of control and unacceptable loads.

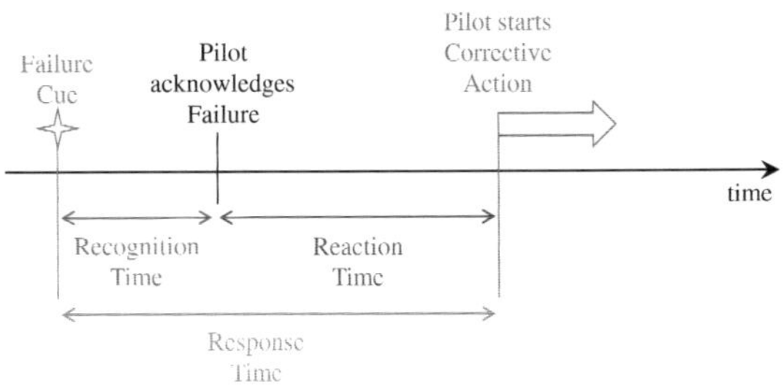

Figure 8 - Definition of Times

Pilots may detect a failure through aircraft motion cues (not applicable at SaIL testing), by cockpit flight instruments or alerts. The specific recognition cues will vary with specific failure type, flight condition, phase of flight and crew duties, [4]: Visual alerts and cues (warning/caution/alert), aural alerts and cues, Physical Cues (at SaIL, environment cues are based on displayed parameters).

Recognition of physical cues in the SaIL is limited because there is no motion base. In general, this led to more conservative results. When the aircraft is disturbed from a steady-state condition, medical science confirms that for the pilot, acceleration cues are processed first by the brain, compared to cues of visual scene change which are registered later. Acceleration cues enable quick control movements to be made in response to transient conditions set by the failure injection. The lack of realistic acceleration cues may lead to flight path deviation due to delayed failure recognition.

The magnitude of an aircraft transient and when it occurs in flight can contribute to the overall safety hazard classification. The transient is a short-term event. *Pilot Response Time* is obtained as the sum of *Failure Recognition Time* and *Pilot Reaction Time*

Pilot Recognition Time is the time necessary to the pilot to acknowledge a failure. It has been estimated to be $T \leq 0.5$ s in the event of visual/aural alert, or physical cue described above.

Pilot Reaction Time is the time elapsed from the recognition of the failure and the beginning of the corrective action. Pilot Reaction Time depends upon the pilot attentiveness at the time the failure occurs. The level of attentiveness depends upon type of recognition cue, the FCS integrity (a degraded system demands a higher attentiveness), flight phase (e.g., take-off/landing, cruise, etc.), Environmental conditions (VFR, IFR).

Verification Criteria

The hazard classification assigned by the FHA (Table 2) is an initial assessment based on assumptions, analysis, and in some cases, test.

Category	No Effect	Minor	Major	Hazardous	Catastrophic
Qualitative Probability	Frequent	Reasonably Probable	Remote	Extremely Remote	Extremely Improbable
Quantitative Probability:	No probability requirement	≤10–3	≤10–5	≤10–7	≤10–9
Effect on rotorcraft	No effect on operational capabilities or safety	Slight reduction in functional capabilities or safety margins	Significant reduction in functional capabilities or safety margin	Large reduction in functional capabilities or safety margins	Loss of rotorcraft
Effect on flight crew	No effect on flight crew	Slight increase in workload which involve crew actions well within crew capabilities such as routine flight plan changes	Physical discomfort or a significant increase in workload or in conditions impairing crew efficiency	Physical distress or excessive workload impairs ability to perform tasks accurately or completely	Fatalities or incapacitation

Table 2 - SAE ARP 4754 Classification of Failure Condition Effects

The effects of the induced failures are judged based on applicable *Pilot Response Time.* The Pass/Fail assessment has a qualitative aspect based upon pilot judgment, and a quantitative aspect based on transient magnitude and ability of the pilot to return to the defined Normal Flight Envelope (NFE). Many factors can contribute to pilot workload and decision making and can influence the hazard classification like: Human Machine Interface (HMI), aircraft handling qualities, emergency procedures effectiveness, crew coordination, environmental awareness. LH developed a structured approach to interview the pilot(s) and address a systematic evaluation of each mentioned objectives which addresses the following points:

- The pilot was asked to give a qualitative assessment of the severity of the malfunction based on Table 2.
- The pilot was asked to evaluate the effectiveness of the emergency procedures, or to propose changes/additions to the emergency procedures
- The pilot was asked if there were any factors in the HMI, cueing, or response of the aircraft that could cause a misleading interpretation of the malfunction.

This testing validates the crew interface with respect to failures by evaluating the pilot's ability to manage aircraft systems and maintain situational awareness. These aspects are a significant contribution to pilot workload. Pilots evaluate if Crew-Alerting System (CAS) messages provide adequate information for decision making. This includes ensuring the highest priority messages are at the top and multiple messages are concise enough that a pilot can understand the situation and decide. Additionally, pilots provide comments on the effectiveness of the displays and cockpit ergonomics.

Since the emergency procedures will link pilot actions to CAS messages and/or audible warnings, pilot comments and data driven analysis could result in a modification of existing EP or in a writing of new procedures. For some malfunctions, a reduction in the flight envelope is required as part of the EP to maintain aircraft controllability – the necessity for this reduction is also evaluated during this testing activity. The impact to mission requirements is assessed to ensure the recommended reduction does not introduce additional hazards (such as reduced range over water).

The pilot's ability to recover the aircraft to the normal flight envelope after the transient is evaluated. The

NFE is based on AW609 pilot experience and expectations for normal operating procedures. Exceedance of the Operational Flight Envelope (OFE) is considered to include a reduction in safety margin and has potential impact on flight loads. However, flight into the Operational Flight Envelope while recovering does not necessarily constitute a reduction in safety and will not exceed flight loads.

The AW609 certification basis combines FAR 14 CFR Parts 25 and 29 (transport category airplanes and transport category rotorcraft, respectively), as well as adding tiltrotor specific requirements, to create the new proposed Part 'TR' (or TiltRotor) rules.

Subpart B of these new proposed rules pertain to flight and flight characteristics. In general, for failure conditions, the controllability, maneuverability, stability, and closed-loop handling qualities requirements of the combined part 'TR' subpart B must be satisfied – if necessary, a more restrictive reduced flight envelope may be defined within which the requirements are satisfied. The aircraft must be able to recover from any transient the failure induced, and transition to this reduced envelope if necessary.

To support the pilot judgment of aircraft response in the Transient Event and during the flight until safe landing, LH defines a minimum set of key parameters that could be extended as necessary, to be monitored and recorded during the simulated flight. A set of parameters to localize the aircraft in its envelope and the engines status during the testing (Table 3), a set to assess the aircraft response (Table 4) and a set to record the CAS messages flowing on A249 protocol. Dedicated data collection is acquired for specific test to verify the proper failure injection.

Conversion Corridor	Hp Vs Mach
ROC Vs Airspeed	Weight Vs xCG @90nacAnd
Roll rate Vs Airspeed	nZ Vs Airspeed
Pitch rate Vs Airspeed	
Left Engine Power vs Airspeed	Left Engine Power vs Airspeed
Left Raw Engine Torque vs ITT	Left Raw Engine Torque vs ITT

Table 3 – Envelope and engine status key parameters

Step_ID and Fail injection time	Actual and Claws Altitude	AoS
Direct Mode Weight On wheels Disc	VCAS, PCAS maxCAS to Display	L/R Rotor Speed
NacAng	L/R Tot. Flapping	nX, nY, nZ

Inceptor Long Tot Long	Inceptor Lat Tot Lat	Inceptor Ped Tot Ped
Pitch Rate	Roll Rate	Yaw Rate
Theta	Phi	Psi

Inceptor Power Lever Tot Power Lever	Ext. Steady State xmsnQ
L/R Torque On Shaft	L/R ITT
	L/R NG

Table 4 – Aircraft key parameters

The following is a summary of a sample test conducted.

Hazard Description:

Erroneous full throttle command to the servo positioning the left or the right engine throttle

Starting Condition:

- Operational phase: Vertical Take Off (Weight OFF Wheels)
- CAS: A/R
- Altitude: field
- Power Setting: TOP
- Nacelle Angle: 90 deg
- NR: 102 %
- Nacelle: static
- Direct/Normal mode: Normal
- Gear up/down: down
- Engine Condition: AEO
- ECL L/R: FLY / FLY
- Turbulence: no
- ROD: N/A
- Trim: N/A
- Note: BEFORE TDP
- GW: LGT (12600)
- SLCG: MM (250.25)
- BLCG: NOMINAL

Test sequence

- Set AC in the Starting Condition and aircraft Configuration
- Start Data Recorder
- Simulate the LEFT Engine runaway by forcing maximum fuel flow to FCU

Description of effects and pilot action:

Aircraft got airborne at time of injection, pilot felt he was gaining vertical speed, then saw an overspeed of NR (103/104 on the MFD rotor speed display). Tried to control the climb by reducing power, while FCS cuts the engine power forcing the pilot to increase power again to compensate for a rapid increase in sink rate. Therefore, he increased power to stabilize the aircraft in a hover then slowly reduce power to gently land.

Note: injection procedure asked pilot to be light-on-wheels to assess worst-case effect.

The following were the list of standardized questions asked to the pilot for each malfunction:

1. *What was the primary indication of failure?*
   a. *Immediate uncommanded rapid vertical acceleration*
2. *Were there opportunities for confusion, miscommunication, misdiagnosis, or errors in procedure that an inexperienced crew would be likely to make and that would make the situation worse?*
   a. *Failure is so rapid that there is no time for communication. No misunderstanding of what is going on but there is a tendency to over counteract the initial climb.*
3. *How significantly was the operational capability of the aircraft impacted?*
   a. *Severe – loss of one engine means flight is scrubbed.*
4. *How significant was the increase in crew workload with respect to coping with adverse operating conditions?*
   a. *No additional increase in workload – transport category takeoff is already a high attentive task*
5. *Were there any aspects beyond increase workload that would impair crew efficiency?*
   a. *No.*
6. *Were the required crew actions within the capability of an average pilot?*
   a. *Yes – especially considering transport category procedure is well established.*
7. *How significantly degraded were safety margins with respect to aircraft controllability?*
   a. *Controllability was not an issue.*
8. *Would pilot recognition and reaction time impact the ability of the pilot to control the aircraft?*
   a. *No – pilot reaction to the failure is immediate in this particular phase.*

SUMMARY OF FINDINGS

The SaIL testing findings could be classified into one of 3 basic categories: simulation-specific deficiencies, misleading or unclear crew alerts, emergency procedure refining. The findings have been resolved with pilot training enhancement, CAS matrix review and emergency procedures refinements. No findings require software fixing or pilot interfaces redesign, demonstrating the high integrity, reliability, and maturity of the AW609 VMS.

The following represent an example of the findings discovered:

Pilot Training and Crew Alerting System optimization

The AW609, being an irreversible control architecture, does not translate control surface forces to the cockpit inceptor. Instead, a spring and motor is utilized to give the pilot a representative force feel. Testing the loss of lateral stick force feel capability, the pilot demonstrated the capability to recover the aircraft after a short period of familiarization with the force feel loss. During the familiarization time frame the pilot noticed a tendency to overcontrol the aircraft on roll axis that could be mitigated with proper review of the 'CYC STICK FEEL' CAS (crew-alerting system) message. The CAS message was previously designated as an "advisory" before the testing. After reviewing the test data, this CAS message has been elevated to "caution". Elevating the CAS to a "caution" created the necessary pilot awareness and attention and reduced the overcontrol tendency in this failure scenario.

The crew-alerting system should use a consistent philosophy for alerting conditions, urgency/prioritization, and presentation. This becomes evident and the design compliance has been demonstrated during the engine anomaly testing ("Erroneous full throttle command to the servo positioning the left or the right engine throttle"). During this test, one engine experiences a rapid and uncontrollable increase in power – the FCS detects the uncommanded increase and rapidly shuts down the affected engine. The CAS messages have been posted in such order that enabled a correct interpretation by the pilot of the hi-side event and informed his decisions for the safe recovery of the aircraft post-failure.

Simulation related issue

AW609 can automatically coordinate turns during appropriated phases of flight. The automatic Turn Coordination (TC) capability reduces pilot workload by allowing feet-off-pedal flight throughout most of the flight envelope.

It becomes clear that having the capability to cue the pilot of lateral acceleration (nY) is a key aspect when the turn coordination is unavailable due to failure. In that case, the pilot shall coordinate the turn maneuver manually. The nY feel deficiencies in a fixed-base platform results in the pilot delayed turn coordination, as the only cue to the pilot is the NY-indicator on the PFD (the 'ball'). The delay represents a conservative aspect in the functional failure hazardous assessment, however having the capability to cue the lateral acceleration via motion could be useful to study the pilot-system interaction during this failure. The missing cue has been successfully mitigated during the testing by informing the pilot, in real time, of the lateral acceleration the aircraft was developing using a dedicated auxiliary display mounted on the pilot side cockpit in a specific position such to not alternate the Human Factor evaluation, but easily viewed by the pilot in the peripheral while maintaining out-the-window visual feedback.

Emergency procedure refinement

Considering the example of the loss of Turn Coordination, the EP was requiring the pilot to coordinate the run manually using the pedal inputs. In airplane mode the coordination requires the system to command a differential collective pitch among the 2 rotors, and this could lead to high rotor flapping situation during the roll maneuver. Considering that the pedal effectiveness is reducing with the airspeed decreasing below a certain value, using the SaIL has been possible to demonstrate that limiting the bank angle with a certain threshold under this failure condition, improves the pilot's capability to coordinate the turn. As such, a reduced, practical flight envelope has been defined for this failure based on limiting bank angle.

LESSONS LEARNED

System malfunction and inappropriate crew response

A highly integrated system incorporates many traditionally independent functions into a single system. Therefore, many system functions, which have typically been separated with limited interdependence, are highly integrated. As a result, there is an increased probability that certain failure modes may now have significant detrimental cascading effects on other important aircraft systems. The pilot can become confused by the numerous CAS messages and might not be sufficiently prompted to mitigate the malfunction which started the cascade. An integrated simulation platform like the SaIL is instrumental in identifying such deficiencies.

Questionnaires help to record the pilot's feedback

The pilot questionnaire that was used for the evaluations was designed to be a structured approach to help the pilot assess the entire failure sequence. Due to the rapid onset and reaction nature of most of the failures, the pilot needs to evaluate multiple factors simultaneously. The questionnaire was used to help the pilot in breaking down the sequence of events. In some cases, in attempting to answer a specific question on the questionnaire, the pilot repeated the failure testing focusing specifically on the specific question. In this way a comprehensive evaluation may be systematically extracted.

SaIL is valid for use as a reasonable simulation of the Type Certificate aircraft

The use of analytical models together with hardware/software integration laboratories (ground-based tools) to facilitate the design development of the Powered Lift vehicle and to partially mitigate company flight test risk has been successfully demonstrated. The relevant SaIL (SaIL hardware, software, articles, or assemblies) specifications which define laboratory such that it is representative of the real aircraft, and Acceptance Test Procedure (ATPs) which verify the laboratory functions like the aircraft, have been conducted to validate the behavior of the laboratory configuration used for the specific tests. When the detailed tests are written by the engineers, they are done so with the understanding of the differences between the SaIL and the aircraft to avoid writing an invalid test.

The regulatory authority has been involved during the development of the SaIL tool. Multiple issue papers have been circulated focusing on topics such as fidelity, configuration control, and acceptability of use, and have been or is being appropriately dispositioned by AWPC. Identification of systems and components that will not be replicated in the SaIL must also be provided. Differences in Item must be substantiated to confirm suitability for use in the SaIL. A report of the ATPs results must be generated and include substantiation of any differences with correlative flight tests (simulator fidelity). The company must provide a process for the scope of re-validation of the SaIL for each change to the SaIL. The process must include a step for substantiation of partial or bypassed re-validation of the SaIL via the ATPs (or regression testing of previously conducted Lab Tests) for each change.

CONCLUSION

SaIL Triple Lab Tie-In (TLTI) has been used to validate the Vehicle Management System (VMS) closed-loop performance and interfaces over a six-month test period.

Findings from this testing led to emergency procedure development and optimization, closed-loop performance of mechanical controls optimization, weight & balance mitigation for the AC#4 RTF, failure mode interaction with redundancy management enhancing and CAS matrix update. Standalone component-specific test methodology would not have highlighted these points discovered, and they would instead have been discovered during flight test - leading to higher risk for the flight test personnel as well as increase resources necessary to facilitate the testing.

The use of representative or actual aircraft hardware ensures a high degree of fidelity in the systems simulation tool to enable early evaluation of the human factors characteristics and closed-loop performances.

The combination of high-fidelity integration giving high confidence in the results, combined with the finding from this sample case, demonstrates that the SaIL is a powerful tool to investigate and discover the most critical condition in which the failure function need to be tested.

An improvement to this real-time evaluation would be a faster-than-real-time batch system to run malfunction tests automatically - this can be used to preliminarily assess the Transient Event without the pilot in the loop. Combined with a robust data evaluation methodology, this could enable the testing of thousands of flight condition combination to explore corner cases that could be hidden to engineering and pilot judgment and experience.

Author contact:

Luca Belluomini
luca.belluomini@leonardocompany.us,
Sid Xiao
sid.xiao@leonardocompany.us

ACKNOWLEDGEMENTS

The authors of this paper would like to acknowledge that this work was made possible, despite the dramatic situation afflicting the world due to COVID-19 situation in 2020, by the efforts of many Leonardo Helicopters engineers, technicians, and managers throughout its 2 plants located in United States and Italy. They would like to especially thank the SaIL Test, Avionics, and Flight Controls teams. The authors would also like to thank the FAA's collaboration to support the closure of the issue paper: "Certification Credit for Failure Mode Testing using Modeling and Simulation Tools".

REFERENCES

[1]. Bartley, G., "Model 777 Primary Flight Control System," *Airliner Magazine*, October–December 1994.

[2]. Aircraft: A Conference on Flying Qualities and Human Factors, San Francisco, CA, January 1993.

[3]. T. Books, D. King, "Hardware-In-The-Loop Simulation (HILS) Testing Of The BA609 Vehicle Management System". Bell Helicopter Textron, Inc. FCS/CS 2002-IV-5-1.

[4]. AC-29-2C Certification of Normal Category Rotorcraft.

[5]. Maisel, M.D., Giulianetti, D.J., Dugan, D.C., "The History of the XV-15 Tilt Rotor Research Aircraft: From Concept to Flight," Monograph 17 NASA SP-2000-4517, 2000.

[6]. Marr, R.L., Willis, T.M., and Churchill, G.B., Flight Control System Development for the XV-15 Tilt Rotor Aircraft," presented at the 32nd Annual National V/STOL Forum of the AHS, Washington, D.C., May 1976.

[7]. Magee, J.P. and Wernicke, K., "XV-15 Tilt Rotor Research Aircraft Program Report," AIAA Paper 79-54, presented at the Atlantic Aeronautical Conference, Williamsburg, Virginia, March 1979.

[8]. SAE ARP4761 - Guidelines And Methods For Conducting The Safety Assessment Process On Civil Airborne Systems And Equipment. December 1, 1996.

[9]. S. Dhanvarachorn, S. Dubeck, K. Schurek, C. Walz, "V-22 Degraded Modes Testing Using a Simulator With Real Flight Control and Mission System Equipment" American Helicopter Society 60th Annual Forum, Baltimore, MD, June 7-10, 2004.

[10]. Kleinhesselink, K., Mason, D., Mitchell, D., and Weakley, J., "Simulation Evaluation of V-22 Degraded Mode Flying Qualities," American Helicopter Society Flight Controls and Crew Systems Design Technical Specialists' Meeting, October 2002.

[11]. FAR Part 25, "Airworthiness Standards: Transport Category Airplanes."

[12]. FAR Part 29, "Airworthiness Standards: Transport Category Rotorcraft."

[13]. Green, D.L., Hart, J., and Hwoshinski, P., "A Perspective on the FAA Approval Process: Integrating Rotorcraft Displays, Controls, and Workload," presented at Piloting Vertical Flight Aircraft: A Conference on Flying Qualities and Human Factors, San Francisco, CA, January 1993.

[14]. McElroy, C.E., "FAA Handling Qualities Assessment: Methodology in Transition," presented at Society of Experimental Test Pilots' Symposium, 1988.

[15]. Fortenbaugh, R.L., King, D.W., Peryea, M.A., and Busi, T., "Flight Control Features of the Bell-Agusta 609 Tiltrotor: A Handling Qualities Perspective," presented at the 25th European Rotorcraft Forum, Rome, Italy, September 1999.

A Missing Piece of PEO Enterprise Architectures: The Assured Infrastructure Modularities

Mark Brown
Mark.Brown1@Collins.com
Principal Architect Collins Aerospace
Minneapolis, MN

Dave Walsh
Dave.Walsh@Collins.com
Associate Director
Collins Aerospace
Huntsville, AL

Scott Nagel
Scott.Nagel@Collins.com
Technical Fellow
Certification & Airworthiness
Collins Aerospace
Cedar Rapids, IA

Harold Tiedeman
Harold.Tiedeman@Collins.com
Technical Fellow
FVL Chief Engineer
Collins Aerospace
Cedar Rapids, IA

ABSTRACT

As the Department of Defense (DoD) pursues the implementation of a Modular Open System Approach (MOSA) to achieve weapon system overmatch through rapid and affordable technology upgrades, a more efficient and effective method to safely certify modularity is needed. In support of the DoD's vision for MOSA, the Assistant Secretary of the Army (Acquisition, Logistics and Technology) (ASA(ALT)) (Ref 1.) has set forth a policy that accelerates the Army's transition from its current document-driven engineering practices to the institutionalized use of modern Digital Engineering (DE) practices, to include Digital Twin and Model-Based Systems Engineering (MBSE). Current DE methods and processes drive a top-down approach to development for functional system specifications derived from enterprise architectures and Family of Systems (FoS) frameworks. These and other efforts intend to satisfy another recent DoD requirement, Modular Open System Approach (MOSA) (Ref. 2). When both DE and MOSA requirements are combined, future complex systems can no longer be developed or certified as monoliths, but rather must be a collection of severable components that can be incrementally certified and deployed as an integrated whole. This paper proposes new enterprise-level governance invariants that can define MOSA components and DE interfaces with sufficient rigor to rapidly certify their modular update. The proposal aims to establish a new coordinate system for identifying onboard modules with required rigor to be useful for both analysis and certification. It aims to promote aircraft sustainability for long lifetimes in the face of modularity drivers ranging across product line, cybersecurity, and safety concerns. The result is a safe, expeditious, consistent, and affordable way to build and certify modular systems that we call Assured Infrastructure Modularities (AIM). It is our intent to submit the AIM for consideration to help define modularity for PEO Aviation's Enterprise Architecture.

CERTIFICATION LACKS MODULARITY

Today, the costs for major weapon system integration and certification greatly exceed the costs of component development. The DoD's ongoing MOSA transformation can be seen as an effort to reduce the costs of system integration and certification because modules are, by definition, easy to separate and integrate. However, an essential question remains on whether the DoD's modularly integrated systems will be easy to certify. We assert that unless a weapon system has a modular architecture and a modular certification approach, its overarching MOSA characteristics will not result in reduced costs of system integration or certification.

A potential solution may be obtained from the DoD's Digital Engineering (DE) transformation. We submit that DE models can be subjected to MOSA queries. Increasingly, DE models provide automated analysis methods which can simulate, analyze, or otherwise predict the success of an integration with respect to important engineering inquiries. However, even excellent DE models lack the information needed to answer queries about MOSA or agile system updates. In some cases, DE models lack any concept of modularly replaceable components. Using existing methods, models lack the MOSA semantics, and do not support modular or incremental certification. To achieve certification, model users must be forced to devote attention to clarify model semantics before a DE model can be trusted for automated certification analyses. Many use cases are being applied across Future Vertical Lift (FVL) programs seeking to achieve business and technical goals, but do the existing MOSA use cases account for an arbitrarily selected change affecting a feature, component, interface, or parameter presumed to be a modular update spiraling into an expensive certification effort? As it stands today, current use of DE models or MOSA components risk not significantly reducing the cost of system integration and certification arising from modular updates. DE solutions of the future must overcome this shortcoming by adding MOSA semantics to models. Our AIM proposal systematically adds MOSA semantics to DE models, as described below.

We propose that enterprises should select the Assured Infrastructure Modularities (AIM) to define an open set of MOSA standards that define levels of modules and their partitioning architectures. While the amount of AIM modularities which best serve each enterprise's MOSA goals can vary, this proposal contains seven modularities which we intend to offer as an initial working set. We also propose establishing invariants for any enterprise that adopts the Assured Infrastructure Modularities. An invariant is defined in mathematics as a function, quantity, or property which remains unchanged when a specified transformation is applied. As such, an invariant of the AIM includes only modularities, that can be integrated and partitioned using a DO-297 Integrated Modular Avionics (IMA) approach. Since each modularity in AIM defines a DO-297 certification path, every AIM modularity will support incremental airworthiness certifications. Since the AIM only includes Open Systems Approach standards, determining whether a candidate module meets the requirements for modular incorporation into an aircraft resolves primarily to verifying its conformance to the relevant open standard. Because each modularity in AIM explicitly requires DO-297 certification of the architectural considerations required to achieve partitioning, each resulting partitioning architecture can be validated and certified to satisfy the isolation otherwise assumed by MBSE specifications' components and interfaces (Ref. 3 & Ref. 4).

AIM modularities can harden DE model semantics and the extent of MOSA, both at time of certification and after deployment, by defining precise boundaries and interfaces for modules. We propose that a certifiable twin of modularity is partitioning. In our proposal, precising boundaries for modules is achieved by selecting a partitioning architecture for each modularity. Seen from another perspective, AIM's users intend to model an aircraft as a nested set of modules, each conforming to a single AIM modularity and its standard interface definition. Using AIM consistently implies that an aircraft is composed of modules, any of which can be replaced. Each of the AIM's modularities define a partitioning architecture as described in DO-178C SECTION 2.4 which affect safety concerns (Ref. 5). For example, considering any of AIM's modularities for software, this statement from DO-178C section 2.4 applies:

"If Partitioning and independence between software components cannot be demonstrated, the software components should be viewed as a single software component when assigning software levels…"

Partitioning architectures decisively alter certification and analysis efforts. We assert they must be explicitly reflected in DE models from the onset of the modeling effort, and not as a late addition to that effort. Historically, the ARINC 653 standard (Ref. 6) provides a case in point. The body of ARINC 653 carefully considers and defines partitioning for a class of partitioning architecture, and then defines a set of Application Programming Interfaces (APIs) which can be implemented by a real-time operating system (RTOS). In retrospect, we observe that the whole of ARINC 653's partitioning architecture considerations are adopted by any enterprise that decides to use that partitioning architecture, e.g., upon procuring its RTOS. Such a decision and its effects provide an exemplar for this AIM proposal, i.e., what we mean by selecting a partitioning architecture for an AIM modularity. While neither DO-178 nor ARINC 653 are considered sufficient to achieve DO-297's modularity benefits, they provide an essential foundation toward that end (Ref. 6). Following the general consideration and definition of partitioning architectures in DO-178C and ARINC 653 as exemplars, we assert that module boundaries can be established by selecting partitioning architectures for inclusion in the AIM. Only after module boundaries are consistently established in the AIM by selecting a partitioning architecture at each modularity will the AIM be able to provide its anticipated MOSA benefits: module reuse among multiple aircraft types, cheap and effective module replacement, and recertification, etc. Based on DO-178C and ARINC 653, we argue that the AIM must be established in aerospace enterprises before modular product line objectives will become cost effective, and before recurring safety and cybersecurity verification and analysis can succeed for a DE model subjected to multiple updates over a lifetime of use.

The AIM requires multiple levels of DO-297 modularity, not just one. Early DO-297 Integrated Modular Avionics (IMA) solutions anticipated only one level of modularity. Now TSO C153A C214 defines a second level of hardware modularity. These compatible, first-generation IMA approaches can host many system functions using shared processing resources called Line Replaceable Units (LRUs) and Line Replaceable Modules (LRM), i.e., processor cards. Since 2009, researchers have anticipated a Second-Generation Modular Avionics (IMA2G). (Ref. 7 & Ref. 8). IMA2G can be seen as a progression from strictly hardware-based and network-based modularities to add software-based modularities. IMA2G's software modularity levels rely on software virtualization to converge critical functions within a logically partitioned, software-defined, IMA-deployed, network computing environment (Ref. 9). IMA2G also entails solutions where software is deployed to multi-core processors, as anticipated by Rushby (Ref. 3) and by CAST-32A (Ref. 10). The AIM collects the levels from both first- and second-generation IMA and is explicit about the open standards used to define each modularity's interfaces and partitioning architectures.

TRANSFORMATIONS THAT WILL CHALLENGE CERTIFIERS

DE transformation is already underway at both the DoD and the world's biggest buyers and sellers in the aerospace industry. Today DE aims to improve the quality, speed, and cost-effectiveness of modeling the final integrated system where each instance may have tailored instances of modules. In contrast to today's Digital Engineering context, prior generations of aircraft afforded a Waterfall Model (Ref. 11) comprised of a linear set of steps used to logically decompose systems into rigidly defined, to-be-integrated functions. As a result, systems using current methods struggle to cope with late-breaking requirements. The Waterfall lifecycle is a top-down and monolithic approach, and large projects that follow it are well known to suffer problems with schedule and cost growth. We need the flexibility to change requirements, verify impacts system-wide, and re-certify. Tomorrow's DE solutions need a way to create a working model of system specifications which can answer business-relevant questions about "will it work?" DE solutions must work well enough to communicate to a team, check type correctness, instantiate and verify for use cases, reify for certification, then integrate system instances following some open, modular approach. We need to know earlier if the draft system specification will work, could work, needs rework, or might work with a new technology or in a new context. DE needs next some way to evaluate and certify the goodness of the model's requirement set, in terms of its safety, cybersecurity and modularity.

This need for system-level model-based evaluation must be held in tension with the increased velocity of requirements changes anticipated by a DE revolution. If DE succeeds, we may expect two important changes to affect system certifiers: 1) DE will enable more complex systems to proceed to certification, and 2) DE will promote draft models to certifiers multiple times for the same system, i.e., requesting preliminary or incremental certification judgments. Given either a Spiral model or an Agile model, certifiers may reasonably expect many more project requests to evaluate a DE model for both safety and cybersecurity problems. Certifiers might expect exponentially more complex future systems, and those requests may be received under conditions where rapid certification responses would be advantageous.

Future certifiers will not decrease their rigor and attention to detail when more details are available from DE methods. Instead, certifiers must remain vigilant to help system owners avoid potential catastrophes. History gives us examples of devastation from Three Mile Island to cyberwarfare such as the SolarWinds Orion supply chain attack (Ref. 12). We now know that greater control and scrutiny must be applied to the requirements specifications for complex systems, i.e., the model side of the Digital Twin. If system builders and certifiers fail to adequately scrutinize complex requirements, future systems may be lost, e.g., "Treat all hosts monitored by the SolarWinds Orion monitoring software as compromised by threat actors and assume that further persistence mechanisms have been deployed." (Ref. 13) Due to the great cost of destruction to complex systems such as nuclear power plants vulnerable to Normal Accidents, and complex military platforms discovered to be vulnerable to cyber-insecurity, this paper suggests that future aerospace systems will need new and standard ways to measure modularity, and their fail-safe, and fail-secure partitioning architectures. For such systems, we as a community need modularity standards which can be objectively tested and certified. As described above for exemplars such as ARINC 653 and DO-178, the modularities must be supported by certifiable measurements of partitioning architectures. Future MOSA governance efforts will be needed to ensure that the modularity standards of the future promote the necessary certification vigilance surveyed here. Specifically, we suggest that future MOSA governance should include efforts to ensure that partitioning architectures selected to support MOSA remain auditable to certain, real constraints in infrastructure which equally constrain specifications as well as DE deployments of Digital Twins.

HISTORICAL ADVICE ABOUT MODULARITY

The DoD has time-tested standards to specify and evaluate procured materiel. However, since Digital Engineering is new, its deliverables may not yet be measurable by or conformant to an adequate set of standards. Similarly, for MOSA, questions may also arise about the selection of the best conformance standards to objectively measure logical or non-physical properties of modularity, cybersecurity, or the propagation of faults through a system. AIM and its governance activities will need to address this gap.

Engineering Management theories since the 1990s sought to characterize different kinds of modularity and their associated modular product architectures (Ref 14.). Initially it seemed that a goal for these studies was to quantify modularity by establishing a single-number answer to measure "the level of modularity." Early efforts sought quantifications similar to McCabe's "cyclomatic complexity" measure which evaluates the in/out degree for a control flow graph of a software function. However, in 2003 authors Mikkola and Gassmann noted a shift away from using theoretically quantified modularity and toward using business drivers to define modularity "strategically." They wrote "Managing Modularity of Product Architectures: Toward an Integrated Theory" reflecting on the actual practices of "firms." These firms are different from aerospace corporations but are also relevant because they managed new product development for product lines within a range of commercial technologies. In summary, Mikkola and Gassmann found that firms can select from among several different kinds of modularity to obtain a custom recipe that promoted their strategy and business goals. Each of the firms observed by the authors of this paper sought to manufacture a "mix" of product models from their product lines to obtain competitive advantage in their market conditions. In their introduction they lay out the motivation for an "Integrated Theory" for engineering management that uses modularity:

"More and more firms (e.g., in consumer electronics, automotive, and elevator industries) are facing difficulties in managing increasing product variety and model mix. A challenge for these firms is to find ways to replace [or reconfigure] products and occasionally, to expand product lines with innovative, high-quality models that minimize development and production costs. Many firms are pursuing modular product architecture design strategies in order to shorten NPD [new product development] lead time, to introduce multiple product models quickly with new product variants at reduced costs, and to introduce many successive versions of the same product line with increased performance levels." (Ref. 15)

Here the authors observe that "many firms" in industry "are pursuing modular product architecture design." Both the difficulties and the benefits described in the quotation above are closely aligned with what the DoD seeks to accomplish today by its MOSA Transformation.

These modularity theorists suggest that in a competitive landscape, winning modular product architecture designs should be selected strategically and not merely by an analysis of some bill of materials or component list. In the case of the DoD's MOSA Transformation, where the DoD can also be seen as "pursuing modular product design," the DoD should strategically identify modularities which best promote its distinctive advantages over competitors. In these case studies, each firm typically faced different market conditions. To illustrate, some use cases required diverse international sales, some required that the top selling model must also be the lowest price model. Some firms leverage internal R&D to create leap-ahead but low-sales-volume models. Rather than a single-number modularity metric, authors conceded: "Component selection embedded in product architectures reflects different strategic choices made by firms." (Ref 15.) Here, "strategic choices" correspond to business strategies adopted in view of market competition and informed by gritty, realistic manufacturing capability assessments. Rather than considering any wishful thinking regarding component modularity, these "strategic choices" should exclude "component selection" that provides low value to the enterprise. Instead, enterprises should strategically reflect upon their competitive landscape and available technology advances to discover useful modularities.

Within a product line context, each manufacturable model is considered a "product configuration." In the case of the DoD, this could equate to a vehicle type or subtype. Given the range of firms observed already using a product line approach, the need for product configurability within the context of a product line is assumed. Instead, authors seek to clarify what is configurable to create a new model within a product line, and more fundamentally, the "purpose" or why product lines benefit from some configurable degrees of modularity. From the Overview, authors advise firms as follows:

> *"Product configurations are rooted in product architecture designs, which may be integral or modular. In assessing modularity of product architectures, issues regarding decomposability (i.e., modularization) as well as bundling of disparate components into a new innovation (i.e., integration) should be taken into consideration. Firms need to understand the fundamental relationships shared between components, respective interfaces, and substitutability of newly developed components embedded in product architectures." (Ref 15.)*

This statement contains a warning that can be applied to DE models: product architecture designs represent a fork in the road, being either integral or modular. MSCs built from those product architectures will perpetuate that decision, becoming either integral or modular. We propose that AIM, an explicitly modular product architecture, is a necessary first step toward obtaining the DoD's intended MOSA Transformation. Even as "Firms need to understand the fundamental relationships shared between components, respective interfaces, and substitutability…" so too must DoD endow its enterprise architectures with this understanding. Without understanding these fundamental MOSA semantics, substituting one component for another won't work. Mikkola and Gassmann provide some helpful case studies to illustrate their point capturing historical attempts at defining modularity, which we quote below in passing.

TABLE I
LITERATURE ON MEASUREMENT OF MODULARITY

Authors	Purpose	Approach and Method
Ulrich & Pearson [49]	• To measure the *manufacturing content* - the attributes of the design that drive cost	• Product archeology - an approach to gather objective data for product development research. Case study: coffee makers
Ulrich et al. [50]	• To estimate the impact of different design alternatives on the net economic benefit of a product	• Economic model to illustrate the relationships among DFM, lead time, and profits. Case study: Polaroid camera
Fisher et al. [17]	• To examine variation in component sharing practice and to identify factors that can explain the variation	• Mathematical model (complemented with optimization, simulation, and regression analyses). Case study: automotive breaking systems
Ulrich & Ellison [47]	• To develop a theory to explain when a firm can benefit from designing product-specific components	• Regression analysis based on survey on engineered, assembled goods. Validation: hypothesis testing with a cross-sectional analysis of 225 products
Collier [8], [9]	• To measure the effect of component standardization on aggregate safety stock levels and service levels	• An analytical measure of product structure termed 'the degree of commonality index'. Tested with simulation experiment

Table 1. Past literature on measuring modularity

TABLE II
ELEMENTS OF PRODUCT ARCHITECTURE

Elements of Product Architecture Modularity	Managerial Implications
Standard components	• Economies of scale • Cost savings • Specialization • Development of capabilities • Standardized interface specifications
NTF components	• Technological risk • Product novelty • Superior performance • Limited imitation • Long NPD lead time • Non-standardized interface specifications
Degree of coupling	• Synergistic specificity • Identification of critical components • Tightness of coupling among components
Substitutability	• Economies of substitution • Component sharing • Product variety • Upgradability

Table 2. Elements of Product Architecture

Above, Table 2 provides four examples of "Elements of Product Architecture Modularity" and their multiple resulting "Managerial Implications." While the use case focus is not relevant to aerospace, it provides a relevant business example of a strategic technology decision. To confirm our case so far, the "Managerial Implications" listed in Table 2 align closely with the DoD's needs for safety, robustness, modularization, technology and innovation management, and competitive marketplace factors, all of which are relevant to our intended aerospace domain.

Later studies (Ref. 16) confirm the necessity of understanding the fundamental relationships between components. Two observations arise and build on the Mikkola and Gassmann case studies and supports the thought process to abandon a single modularity metric akin to McCabe complexity. Furthermore, that aspects of business strategy, intelligent selection, and components' operational and manufacturing fundamentals should be synthesized to identify cost-effective "degrees of modularity":

"*Modularity, modular product architectures, and modular product platforms have recently gained popularity in product development. Modular architectures, consisting of loosely coupled chunks (modules), have many benefits from cost savings due to commonality to independent design of modules. However, a fully modular design may not always be achievable when designing complex systems and products. It is suggested that modularity is not simply a binary characteristic, but that products can and do exhibit varying degrees of modularity. In this study, the authors first develop a new measure of modularity and then show that designs that are heavily driven by weight, size, or other performance constraints, often exhibit rather integral architectures [not-modular architectures] commensurate with a smaller degree of modularity. In contrast, when the design is driven mainly by business goals, such as cost savings or commonality, a more modular architecture often emerges. This is shown by quantitatively analyzing the degree of modularity of existing product architectures and qualitatively by discussing the relationship between technical and business constraints." (Ref 16.)*

In summary, we should anticipate many kinds of modularity because "modularity is not simply a binary characteristic." Authors here advise that "a more modular architecture often emerges" when an enterprise focuses on "business goals" instead of size, weight and power.

Given this history and advice concerning modularity, our proposal to collect Assured Infrastructure Modularities seems quite fundamental. We have seen that not only has DO-297 defined multiple levels of hardware modularity, but IMA2G adds levels of software modularity. We see that DoD intends both MOSA transformation, to achieve modularity, and also DE transformation, to utilize digital models and enterprise product architectures over the course of product lifetimes. Literature confirms that several kinds of modularity can be expected, and further, that modularities should be chosen carefully and strategically. We have received the advice that modularities should be strategically selected for how well they promote the business goals: "In contrast, when the [modularity engineering effort] is driven mainly by business goals, such as cost savings or commonality, a more modular architecture often emerges." This is confirmed by the DoD in its 2013 OSA Contract Guidebook for Program Managers. Specifically, the essence of OSA is organized decomposition, using carefully defined execution boundaries, layered onto a framework of software and hardware shared services and a vibrant business model that facilitates competition.

ASSURED INFRASTRUCTURE MODULARITIES

We now turn to framing some initial modularity standards in AIM that can help us achieve the DoD's business goals. We seek initially to reduce the cost of re-certifying a future modular substitution of hardware or software. Our scope includes certification of both cybersecurity and airworthiness. Given our explicit IMA certification goals, we propose the following modularities for the AIM:

- (Hardware) a **LRU** box as a module | Ethernet as a partitioning architecture, e.g., network protocols for network packetization and addressing; may infer encryption and rate controls
- (Hardware) a **LRM** card as a module | Ethernet as a partitioning architecture
- (Hardware) a **Core** as a module | CAST-32A as a partitioning architecture
- (Software) a **Virtual Machine** as a module | Hypervisor as a partitioning architecture
- (Software) a real-time **Partition** as a module | Kernel-based partitioning architecture, e.g., ARINC-653 RTOS time and space partitioning
- (Software) an **Application** as a module | Kernel system calls for message/file partitioning
- (Data) a typed **Message** as a module | Message type store-and-forward, routing, or direct links as a partitioning architecture, e.g., FACE TSS message routing.

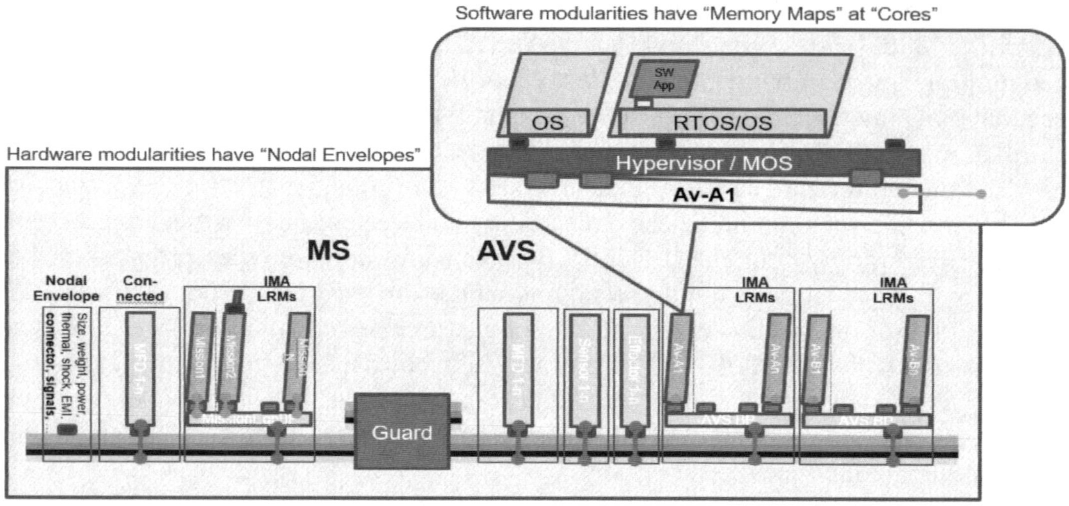

Fig. 1. Hardware and Software modularities contained in the Assured Infrastructure Model (AIM)

The figure above shows physical hardware infrastructure in the red-outlined rectangle; both large avionics boxes and small avionics sensors are shown plugged into the same physical network layer, which together comprise the first level of modularity (LRU). Some avionics boxes shown include a modularity bus called the "backplane" for the box, into which multiple cards may be inserted, which together comprise the second level of modularity (LRM). Still within the realm of physical hardware but at a microscopic level, each card may provide a multicore processor, one or more DMA engines, co-located graphics, floating point, machine learning, or other hardware "IP core". When an IP core is connected to a shared coherent memory fabric, we say that the cores and that fabric comprise the third level of hardware modularity (Core). AIM users say that each module "pops" into its associated modbus at the same level of modularity, emphasizing a fundamental metric of modularity, i.e., modular attachment to the Assured Infrastructure.

Also shown in the diagram above are three modular levels of software assured infrastructure located within the red-outlined rounded rectangular callout. We sketch AIM's generic model for its hardware-to-software interface as follows. Generically, software executes on a hardware Core after registers in that core are set to values which define and enable the software context or environment to run. Unconfigured hardware cores are depicted in Fig. 1 as gray rectangles located below the hypervisor, and contrast with the register-configured cores projected through the hypervisor. Hypervisor-configured cores are shown as the small black rectangles located above the hypervisor layer. Similarly, a kernel or "OS" can be used to configure even more registers in a core and project a more-configured core to its modules. Each AIM software level contains a memory partitioning architecture that defines its own "memory map". A memory map not only can partition memory, but within each partition's memory space it might also aggregate all the software-accessible memory and hardware registers into a set of contiguous memory regions. In AIM, each successive software level creates finer-grained partitions of memory and provides a single memory partition for the exclusive use of that level's partitioned software. Each of the memory maps for modules at the software levels are stored in memory, so we assert that so long as all memory mappings are known to an auditor, these software partitioning can be objectively measured.

The "software" AIM levels also provide partitioning and modularity. The hypervisor provides a modbus at the hypervisor domain and hosts multiple fourth-level software modules (Virtual Machine instances). The supervisor/operating system kernel separates time, space, and IO device access for each software container (partition). The application syscall Application Binary Interface allows one or more software functions to invoke privileged routines in a shared kernel. This provides the basic concept of software partitioning in AIM, though additional considerations apply. The next level is data modularity. The set of all functional message types which are defined by the enterprise. While at first glance, this modularity may seem to include all possible network data transmissions, for today's networks comprised of multiple layers of networking protocols, it does not include many potentially important lower-level network protocols and their transmissions.

For brevity, we omit discussion of network protocols, i.e., for host and equipment administration, authentication, encryption, time synchronization, time partitioning, etc. The AIM proposal anticipates its users will consider selecting additional modularities, e.g., relating to data at rest ("storage") and data in transit ("network"), as shown below. As illustrated in Fig. 2, below, AIM can be extended to more levels following the AIM pattern

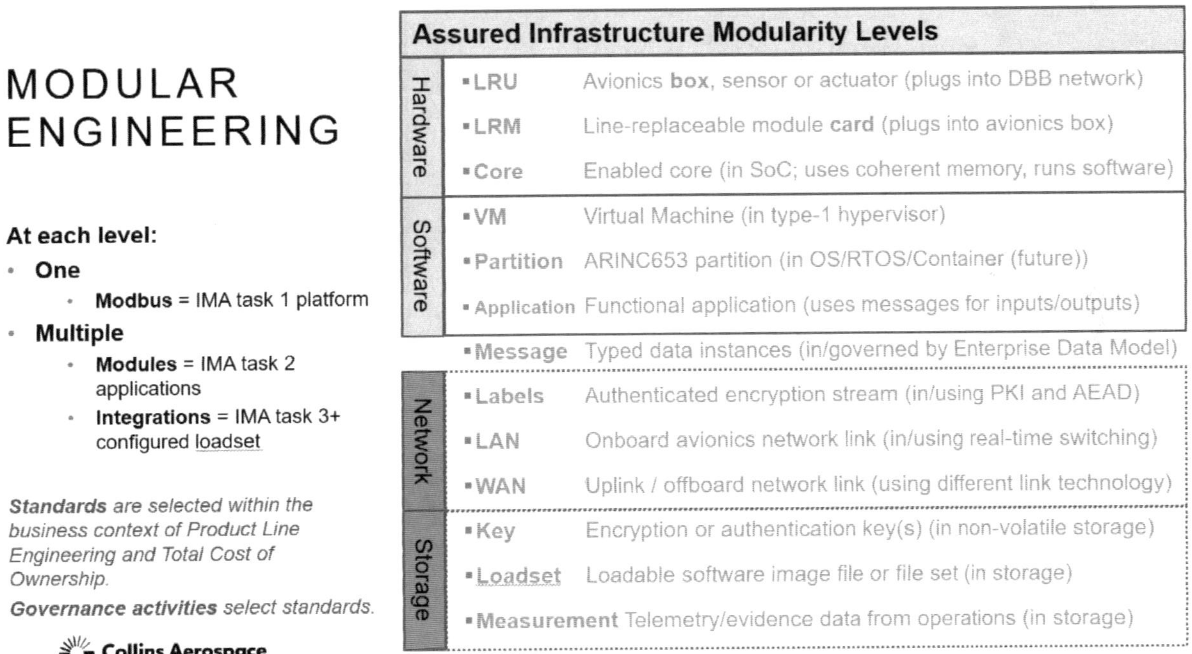

Fig. 2. Additional Candidates in the Assured Infrastructure Model (AIM)

Taken together, these hardware and software modularities intend to define an objectively measurable and physically verifiable coordinate system for future deployment of hardware and software modules to a vehicle or other local area network.

GOVERNANCE FOR MODULAR ARCHITECTURE STANDARDS

The AIM set is open and extensible to future modularities being defined by its users when the following conditions hold for each candidate modularity:

- **Assured partitioning infrastructure**: the modularity can be established by deploying a partitioning architecture capable of hosting one or more modules and being certified under DO-297.
- **Standard connector type(s)**: the proposed AIM level identifies one or more standard connector interface type(s) at which points modules are allowed to connect to the infrastructure.
- **Interface governance process:** governance can be established capable of revising the standard(s) used at the module connectors for the level.
- **Replacement governance process:** governance can be established capable of replacing the level's partitioning infrastructure, i.e., at a higher level in AIM.

We propose that each enterprise should provide an ongoing governance service for their Assured Infrastructure Modularities to handle future shifts in standards over time. While each enterprise's governance decisions are outside the scope of this paper, in the following we present an overview of examples to illustrate the governance efforts that are appropriate for the AIM.

First, referring to the conditions that must hold for any candidate modularity within the AIM, governance must evaluate if future proposed candidate modularities or standard update meets the AIM conditions. For an example, consider an enterprise that has already established a set of standards for AIM at the "(Hardware) LRU box as a module | Ethernet as a partitioning architecture." Following the description above, we grant that the enterprise in our example has already selected and deployed a specific version of IEEE 802.1 and related Ethernet standards as the partitioning infrastructure sometimes referred to as the Digital Back Bone (DBB). Whereas "Ethernet" refers to a large family of standards, here "DBB" refers to a subset of standards selected for inclusion by the enterprise, qualified for airworthiness, and additionally subjected to a DO-297 incremental certification regime at the first level of AIM, i.e., providing the IMA Platform at the LRU.

In this example, we imagine that the enterprise governance task is to consider a new minor standard change to be included in its LRU level. The new standard could be any future standard that purports to be within a relevant family of standards, i.e., it could be a standard that revises, updates, or extends either the DBB's IMA Platform or the DBB's standard connector type for modules. Since the DBB at the LRU level must remain open, following an Open Systems Approach, the enterprise governance must give preference for open standards, but must also consider draft open standards as the presumptive next generation. Some examples of Ethernet standards which could be relevant to a DoD enterprise's DBB at the LRU level include: Copper vs. optical fiber Ethernet media; Different standard Ethernet "PHY" signaling rates, e.g., measured in megabits, gigabits or terabits per second; different standards for Ethernet "MAC" interface (wiring) approaches; additional networking protocols integrated with the Ethernet "MAC" layer, e.g., network time synchronization, encryption, rate limiting or time triggered protocols, etc. For each of these historical examples, it could be predicted that future revisions of that standard might appear on the horizon and require a governance decision. The governance decision must not only consider the cost to improve the IMA Platform, likely an Ethernet switch and attached transmission media, but the impact of any change in the LRU level standard suite to its attached module constituents across all of the enterprise's deployments.

In a second example, we imagine that the enterprise governance task is to consider a major standard change, e.g., a request to accommodate a non-Ethernet network for onboard communications. The requested standard could be a legacy avionics bus or network, for example a serial bus or a military standard. The governance task may include triaging the request by discovering if a network translator could be used, i.e., Ethernet to legacy bus, which might depend upon data transmission performance and latency requirements and available COTS/GOTS products. On the other hand, a major standard change request may suggest a novel or leading-edge technology standard, for example CXL (Ref. 17), the Compute Express Link, a way to post and receive cache-coherent memory transactions from one chip (package) to another using PCI Express data links (version 5.0 and above). In the case of CXL, governance activities might investigate the suitability of CXL for use in onboard avionics networks and prepare a cost-benefit analysis for introducing CXL to the fleet if warranted.

In a third example, we assume further that CXL governance activities resulted in a go-forward decision at the enterprise. We still assume that an 802.1 Ethernet family of standards has already been deployed across the enterprise fleet over the course of many years, that several minor standard updates have been selected and procured over that time, and further, that a variety of harmonized Ethernet standards suites are now currently deployed to subsets of the enterprise fleet. Additional governance activities might identify the AIM level for which this new link was most suitable and prepare a transition plan. If, for example, governance activities found that CXL was best suited at the AIM, "(Hardware) a LRM card as a module | Ethernet as a partitioning architecture," and not at the LRU level, we can observe that simply *making this governance decision* would significantly aid the technical work of deployment planning. Specifically, the media deployments which are used to carry LRU level transmissions (i.e., relatively long runs of onboard communications between boxes, sensors, etc.) can be observed to have different attributes and different replacement costs compared to the communications media runs deployed at LRM level in the enterprise (i.e., relatively short runs used to connect cards within the same avionics box).

Given this third example, where an enterprise elected to add CXL to the LRM level, how can we anticipate its costs and benefits, especially with respect to MOSA goals? First, we have already assumed that the enterprise governance activity has compared its legacy Ethernet solution at the LRM level to its new CXL solution. We have assumed that the enterprise governance activity has created a list of costs and benefits for updating AIM to CXL at the LRM level. The costs will certainly include procurement of a CXL solution for at least one avionics box, i.e., where the modules (LRMs) in that avionics box (LRU) have been upgraded from using an Ethernet backplane to using a CXL backplane. In this example, we can see that each enterprise's governance activity should decide whether backward compatibility with legacy Ethernet is required for the IMA Platform at the LRM, e.g., using a physical plug-in connector capable of both Ethernet and CXL (PCI express), or whether it is more cost-effective for the enterprise to allow all of the avionics boxes (LRUs) in its fleet to carry either legacy Ethernet or novel CXL across its backplane at the LRM level. The enterprise should consider obvious benefits by directly comparing the impact of using the CXL protocol versus the legacy Ethernet protocol. Additionally, the enterprise may consider deployment project costs and schedule, e.g., observing that longer rollout times often result in greater costs and a greater number of well-defined and verified intermediate steps toward project completion. Doing so, the enterprise may consider a range of deployment options. It should forecast a future state when all card modules participate only in the upgraded CXL standard for communications. Also, the enterprise governance activity will likely forecast how many of its future aircraft functional modules will require the benefits of CXL, and how valuable those future aircraft functions are to the enterprise versus legacy functionality which does not require CXL. These governance activities can be repeatedly applied to any single level of AIM, for any MOSA proposal about a new or updated open standard that the enterprise anticipates will impinge upon its set of Assured Infrastructure Modularities. Below, we consider more generally the MOSA benefits due to AIM, extending from this specific example of how to reap MOSA benefits after a governance decision has been made to deploy a new major standard (CXL) targeting a specific AIM level.

APPLYING IMA TO MULTIPLE LEVELS

This AIM proposal anticipates reduced integration and certification costs due in part to applying DO-297 IMA at multiple levels. To support of our claim that IMA levels may be nested we offer the following explanation. The figure below is quoted from DO-297, and it provides an overview of the chapters within that document.

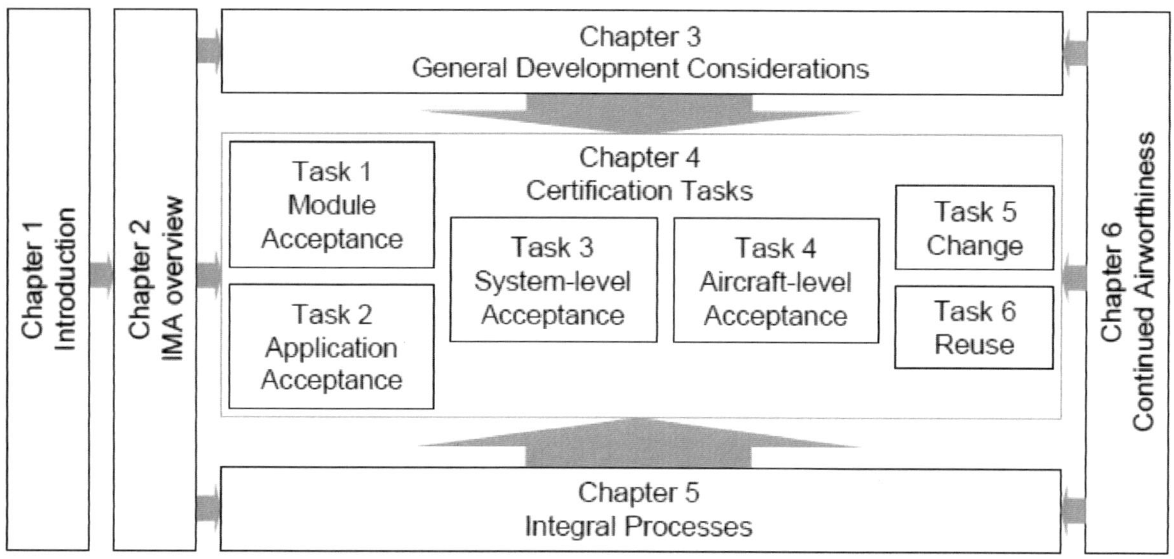

Fig. 3. DO-297 Integrated Modular Avionics (IMA) chapters and their relationships

At the center of the figure lies a box representing "Chapter 4 Certification Tasks" which encloses the six IMA "Tasks". Since the figure above only shows one level of IMA, a question may arise about whether AIM's multi-level IMA can be conformant with DO-297?

To support our claim, consider the IMA terms defined in the following quotation from section 1.2 "Scope" in DO-297:

> *As defined in this document, IMA is described as a shared set of flexible, reusable, and interoperable hardware and software resources that, when integrated, form a platform that provides services, designed and verified to a defined set of requirements, to host applications performing aircraft functions." (DO-297, p. 2)*

Unpacking this complex IMA verbiage, we summarize it to say that on the one hand there is the IMA platform, and on the other hand the platform will "host applications performing aircraft functions." A major thrust of DO-297 describes efforts to "form an [IMA] platform," which is accomplished by IMA Task 1. Using the AIM, DO-297 Task 1 must be repeated once per AIM level, i.e., AIM requires that an IMA platform be established at every level in order to host aircraft functions deployed as AIM modules. Our claim that IMA levels may be nested arises from the fact that IMA platforms established in Task 1 are either/both "module" and/or a "platform". For example, in Table 3, Task 1 is listed as this Integration Activity: "Integrate components and/or modules to form a platform." In the same row, the certification Task is listed as "Module and/or platform acceptance." This means that IMATask 1 anticipates the possibility of integrating IMA modules into an IMA Platform. We assert that this could be used to nest IMA Platforms.

This ambiguous wording in DO-297 ("module and/or platform") is necessitated when an LRU box forms an IMA platform, i.e., when an LRU accepts LRM cards, and the LRU itself may be modularly replaced because the aircraft and its network provides another IMA platform. Nested modules can be seen in IMA TSOs C153a (Ref. 21) and C214 (Ref. 22). These recent IMA TSOs explicitly define IMA with multiple levels of hardware (Rack and LRM). In TSO C153a, it defines IMA Module as "A component or collection of components that may be hardware or a combination of hardware and software, which provides resources to the IMA-hosted applications…." In a complementary analysis, we can see that "applications" in DO-297 may be hardware-with-software modules, for example a LRM card (e.g., in "IMA Design Terminology" in DO-297 section 2.1.1, Aircraft Function is defined to include "the capability of the aircraft that may be provided by the hardware and software of the systems on the aircraft." p.7) This can be confirmed as follows. In IMA both the LRU (box) and the LRM (card) may be implemented as a composite of hardware and software, and both may contain an aircraft function, i.e., per DO-297's definition. In contrast to the platform/module ambiguity, the following term in IMA seems straightforward: "Functions may include flight control, autopilot, braking, fuel management, flight instruments, etc." (p.7) We assert that whether a particular aircraft function is deployed in a larger format e.g., as an LRU, or in a smaller format e.g., as an LRM, is a matter of indifference in DO-297 and its stipulations. This explains why Task 1 results in certifying "Module and/or platform acceptance" in DO-297. The primary concern is whether an aircraft function is *hosted by an IMA platform*; that makes it modular and amenable to IMA Tasks 2-6, summarized below. Whether that aircraft function can be X-rayed and shown to contain any nested IMA platform with its own set of modules is in comparison a secondary concern in DO-297.

We observe that the balance of the DO-297 describes efforts to integrate hosted applications and certify an IMA Platform into a larger unit called an "IMA System." DO-297 scope includes both the initial task to "form a platform" and the remaining modularity tasks needed to "host … aircraft functions." That is, the additional DO-297 "tasks" include their per-application certification (Task 2), their integration with the platform and that IMA System level certification (Task 3), their collective certification up to the level of aircraft certification (Task 4), plus specific callouts for change (Task 5) and reuse (Task 6). Given even this cursory review of the certification tasks in IMA, we can observe that reduced cost and risk for integration and certification, i.e., the stated goals of MOSA transformation and the goals of AIM, should be satisfied by DO-297. We admit that each enterprise's instantiation of DO-297 IMA may vary in terms of how well it achieves the benefits of reduced cost and risk. However, it seems clear that the goals of DO-297 align very well with the goals of MOSA transformation, such that no other avionics certification standard can better serve MOSA goals.

MOSA BENEFITS FROM AIM

Our AIM proposal appears to align with the overarching goals of MOSA transformation, and secondarily, be informed by DE transformation. As stated earlier, the AIM and its governance must answer MOSA queries such as, "will arbitrarily selected change affecting a feature, component, interface, or parameter presumed to be a modular update spiral into an expensive certification effort?" Integration, certification, and fleet-wide deployment have historically been inefficient, costly, and these concerns now motivate a MOSA transformation. We now propose how the AIM might serve the enterprise on those points.

MOSA benefits can be realized in the third example provided above, which shows how a single level of assured infrastructure, already certified as an IMA platform, can be modularly replaced in the AIM. Whether by replacing an entire modularity represented by an AIM level, or by replacing any single module or modules, vendor lock can be broken. While the wholesale replacement of an IMA platform within the enterprise fleet as illustrated in this example represents one of the greatest MOSA challenges that the AIM can satisfy, next we consider whether the CXL example can be generalized by the AIM. Informally, the CXL at the LRM level fleet deployment example selects a novel avionics network within our anticipated enterprise context, that suggests that it is not a unique case or somehow chosen to exaggerate the AIM's benefits. That is, since CXL has never been deployed to an aircraft before, nor has it been certified with respect to airworthiness, cybersecurity, or otherwise, it can be considered a fresh example that poses a realistic challenge to the AIM's invariants. In the following analysis, we intend to justify speaking generally about the AIM, i.e., about the limits and capabilities that the AIM conveys to its enterprise based on its invariants. We can follow this advice given above:

> *"In assessing modularity of product architectures, issues regarding decomposability (i.e., modularization) as well as bundling of disparate components into a new innovation (i.e., integration) should be taken into consideration." (Ref. 15)*

We take CXL to represent a "new innovation" which could be used either to replace an existing Ethernet modbus at the LRM level, or potentially could create a new AIM level altogether, for example, by introducing a new modbus between chip packages soldered onto a single LRM card. For such cases of novel standards and deployments, as well as for lesser changes, we assert that the AIM's invariants remain constant and remain useful to the enterprise. Again, the definition of an invariant is defined as a function, quantity, or property which remains unchanged when a specified transformation is applied. In the outline given below, we re-organized our initial statements about the AIM's invariants into three main assertions:

1. **The AIM only includes Open Systems Approach standards.** Determining whether a candidate module meets the requirements for modular incorporation into an aircraft resolves primarily to verifying its conformance to the relevant open standard.
2. **Every AIM level supports incremental airworthiness certifications.**
    a. AIM includes only modularities which can be integrated, partitioned, and certified using a DO-297 Integrated Modular Avionics (IMA) approach.
    b. AIM's users intend to model an aircraft as a nested set of modules, each conforming to a single AIM level and its standard interface definition.
    c. Using AIM consistently implies that an aircraft is *only* composed of modules, any of which can be replaced.

3. **Every AIM level selects a partitioning architecture which affects certified concerns**, as described in DO-178C section 2.4.
    a. Because partitioning architectures decisively alter certification and analysis efforts, we assert they must be explicitly reflected in DE models.
    b. Module boundaries can be defined using partitioning architectures included in AIM.
    c. AIM DE models harden model semantics and the extent of MOSA, both at time of certification and after deployment, by defining precise boundaries and interfaces for modules at each AIM level.
    d. Applying cybersecurity analyses to AIM-based aircraft, including STPA (Ref. 18) or cybersecurity resilience analyses, can also benefit from its precisely defined module boundaries and their standard interfaces.

Since the organization presents a fundamental challenge facing those who would improve the modularity exhibited by their enterprise or technical systems, we believe the organization of the statements given above should be critically evaluated. Does this organization, above, reducing our proposed invariants governing the Assured Infrastructure Modularities to three main assertions, promise help enterprises? We consider if the AIM's three main invariants provide essential MOSA benefits, below.

The first main invariant is that the AIM only includes Open Systems Approach standards. We intend this assertion to be interpreted *ex post facto* with respect to whether the AIM already contains a set of modularities arranged in levels. The first main assertion states a simple invariant constraining the set of standards in the AIM, past, present, or future. While we admit some ambiguity about whether standards are "open" or not, we entrust these concerns to the enterprise's governance function. Granted the historical advice about the strategic nature of modularity decisions presented above, our emphasis is that not only must each AIM standard be selected for its openness, but also it must be carefully and strategically critiqued for its modularity advantages to the enterprise. Identifying modularity standards which are useful to an enterprise's strategic goals is more difficult than merely identifying open standards. Specifically, we have suggested that an appropriate AIM governance function must "prepare a cost-benefit analysis for introducing [any new MOSA standard] to the fleet if warranted." While making any strategic decision for an enterprise should be considered on face value to represent a challenging activity, our literature review supports this claim. For those who have not yet invested in reading its referenced body of knowledge and yet seek to obtain enterprise benefits from modularity, we commend the exercise given the following caveat shared by Mikkola and Gassmann in 2003 (Ref. 15):

> *Many studies on modularity are qualitative and exploratory in nature (cf. [2], [5], [19], [25], [28], [37], and [41]). ... Although these studies have contributed greatly to our understanding of modularity, they offer limited insights into how firms measure the degree of modularity embedded in product architectures. Nevertheless, Table I lists several studies that focus on measuring modularity.*

Our second main invariant is that each AIM level must support incremental and modular airworthiness certifications, implying DO-297 IMA or its successors. We have shown that DO-297 does not necessitate a single level of IMA modularity. We have argued that the primary concern of DO-297 is whether an aircraft function can be hosted by an IMA Platform, such that is amenable to Tasks 2-6. DO-297 provides a rigorous and detailed certification path to establishing multiple kinds of modularity. We have explained our interpretation about the Scope of DO-297 that, in summary, a successful completion of Task 1 overcomes the challenge of producing a modular infrastructure, and given that success, all subsequent Tasks 2-6 can be expected to deal with its compatible modules. While we admit that overcoming the challenge of defining a candidate modularity to the level of rigor needed to successfully complete Task 1 in DO-297, i.e., such that Tasks 2-6 can then be completed straightforwardly, multiple aerospace enterprises can produce historical examples of this kind of success. In other work, authors of this paper have investigated the possibility of certifying IMA2G by following the guidance of CAST-32A in more detail. CAST-32A suggests that its definition of "robust partitioning" quoted from DO-297 provides a key indicator for whether a multi-core processor can modularly host aircraft functionality in the form of software applications. While "robust partitioning" is not a sufficient condition for establishing a DO-297 IMA Platform, it is necessary, and its necessity traces back to the earliest efforts and publications in the field of aircraft certification, e.g. (Ref 3, 19).

The third main invariant emphasizes the non-functional requirements for partitioning (or similar terms including separation, isolation, non-interference or independence). At every level of the AIM, modules must be partitioned from each other using a partitioning architecture. We admit that imperfections in a partitioning architecture called interference channels can pose thorny conundrums (Ref 20). However, in the context of aerospace certifications for real-time deadlines and verified software requirements, timing channels and covert channels will be subjected to systematic mitigations. Even for hardware interference channels present on multi-core processors, CAST-32A anticipates that mitigations may be found and certified within the context of a "safety net" (Ref. 10, e.g., objective "MCP_Resource_Usage_3"). For airworthiness certification, the partitioning architectures which govern AIM modules must be adequate to contain failures within the larger context of common cause analysis. We believe that the AIM's proposal to introduce diverse partitioning architectures at each of its levels could potentially reduce the likelihood of failure propagations, especially when the strengths, weaknesses and distinctives for each level's partitioning architecture are critically evaluated. We suggest that this concern should be addressed during DO-297's Task 4, "Aircraft integration," emphasizing both the separation requirements and common cause analyses which should vest at this level of certification.

Our proposal emphasizes AIM governance that intends to establish a set of modularities according to open standards, partitioning architectures, and DO-297 certification rules. To evaluate the AIM for its MOSA benefits as intended, an enterprise must also task its accounting and management functions to evaluate the AIM's benefits in terms of cost and schedule. Specifically, for each of the AIM levels the enterprise must be able to estimate the cost of the modbus and module procurement, integration, certification, and deployment to its fleet. The cost and schedule benefits of functional/non-functional allocations to specific MSCs modularity must also be evaluated for enterprise reuse and product configuration options. For these tasks, enterprise management should consider obtaining both estimate and actual costs and durations for each of the six tasks of DO-297 for each level in AIM.

SUMMARY CONCLUSIONS

We proposed three invariants to guide enterprise MOSA governance activities while illustrating these invariants across seven levels of modularity. By following these invariants, module boundaries can be consistently established and MOSA governance can be predicted to yield MOSA's anticipated benefits: module reuse among multiple aircraft types, and affordable, effective module replacement, integration, and recertification. The invariants can prevent vendor lock, especially when subjected to the recommended governance activities. To implement the AIM, governance activities should start by continuously identifying new, open technical standards. At the same time, governance activities should strategically seek to identify modularities which may emerge after careful review of competitive factors, standards, and total cost of ownership concerns including the cost of certification. Our second invariant directly requires a consistent DO-297 regime to enable modular avionics certifications. Our third invariant aligns each enterprise's selected modularities with more foundational concerns about partitioning versus common cause analysis and whether a partitioning architecture will be sufficient to prevent fault propagation. When an enterprise defines a set of modularities, follows these invariants, and uses its AIM levels for procurement, integration, certification, deployment, and cost accounting, it has embraced this AIM proposal.

The AIM follows the DoD definition of OSA, and OSA-organized modularity defines the kinds of systems we will need to certify in the future. To reiterate:

> *"The essence of OSA is organized decomposition, using carefully defined execution boundaries, layered onto a framework of software and hardware shared services and a vibrant business model that facilitates competition."*

The AIM promotes this and emphasizes carefully defined execution boundaries that carry incremental certification benefits under the DO-297 umbrella. The AIM is rich enough to include concurrent execution, parallel execution on multi-core processors, and distributed execution while maintaining cybersecurity controls (control vs. data plane, multiple levels of security and encryption) and safety controls relevant to real-time execution deadlines. The AIM defines a MOSA that allows for modular and incremental certification after software module replacement, supporting common business goals and strategies including facilitating competition for future software module updates and expansion of the infrastructure to support more software and hardware modules deployed to the same aerospace platform. The reasons behind the AIM's invariants that promote modularly certifiable systems include not only business strategy and cost-effectiveness, but also apply past lessons learned from the history of modularity.

In summary, the AIM identifies modular certification and its governance as a missing piece; this deficiency presents an obstacle to MOSA adoption in DoD enterprises. Enterprise management and its governance activities should place new emphasis on procuring partitioning architectures, certifying IMA Platforms, and continuously seeking updated open standards. Without over constraining the set of modularities any enterprise might need to meet its strategic and business goals, the AIM provides a set of invariants that, when promoted by a governance activity, can surely achieve the DoD's MOSA transformation goals. Following the AIM invariants can lead to effective and efficient replacement of hardware, software, and other infrastructure levels which can be subjected to standards, AIM governance, and certification. These qualities are fundamental to supporting the overarching goals of DoD MOSA to achieve weapon system overmatch through rapid and affordable technology upgrades.

FUTURE RECOMMENDED WORK

As explained throughout this paper, the AIM provides a consistent template to establish nested IMA platform and partitioning architecture considering the larger system context. This critical infrastructure satisfies PEO Aviation goals of creating reusable groupings of capability that can be incrementally certified. The AIM starts with the traditional LRU, to LRM reusability, and with the implementation of the IMA2G, an Operating Environment inclusive of the hypervisor, RTOS, and its applications.

To demonstrate the benefits realized by the AIM, we recommend future collaboration with the US Army MOSA Transformation office to perform an integration with the components within the Enterprise Product Architecture (EPA) that handle navigation in commercial airspace.

ACKNOWLEDGMENTS

This paper was developed through a series of meetings with specific content and perspectives provided by the following individuals in addition to the authors: Matt Sipe, Scott Dennis, John Stough, Dr. Tom DuBois, Keith Zook, Pat Gomez, Tom von Eschenbach, Jonathan Demildt, Tony Adams, Dave Brown, Dave Radack, Chris Noll, and Dan Kaplan.

REFERENCES

1. Jette,B.D, “Policy for the Implementation of Digital Engineering Throughout the Assistant Secretary of the Army (Acquisition, Logistics and Technology) Enterprise,” ASA(ALT), 11 April 2022

2. United States. Congress. House. Committee on Armed Services. (2016). National Defense Authorization Act for Fiscal Year 2017 : report of the Committee on Armed Services, House of Representatives on H.R. 2953 together with dissenting views (including cost estimate of the Congressional Budget Office). Washington: U.S. Government Publishing Office

3. Rushby, J, “Partitioning for avionics architectures: Requirements, mechanisms, and assurance. NASA Contractor Report CR-1999-209347, NASA Langley Research Center, June 1999

4. Rushby, J, “Modular Certification,” Technical report, Computer Science Laboratory (CSL), SRI International, Menlo Park, CA, September 2001

5. DO-178C Software Considerations in Airborne Systems and Equipment Certification. RTCA, December 2011

6. ARINC 653 Avionics Application Software Standard Interface, November 2010

7. Fuchsen, Rudolf. (2009). Preparing the next generation of IMA: A new technology for the scarlett program. 7.B.5-1 . 10.1109/DASC.2009.5347424. See also https://acq.nl/scarlett/

8. Sheynin, Yu & Suvorova, Elena & Bukov, Valentin & Shurman, Vladimir. (2014). Integrated Onboard Networking for IMA2G. 10.13140/2.1.1190.6569. See also https://cordis.europa.eu/project/id/605442

9. Dova, V.H, "Software-defined avionics and mission systems in future vertical life aircraft," Calhoun: The NPS Institutional Archive, http://hdl.handle.net/10945/45181, March 2015

10. CAST-32A. Federal Aviation Administration. 1 November 2016

11. Barry Boehm, "A spiral model of software development and enhancement," IEEE Computer 21:2 (May 1988)

12. DHS Emergency Directive 21-01

13. DHS ED 21-01 4.a

14. Ulrich K. T., "The role of product architecture in the manufacturing firm", Research Policy, 24, 1995, pp. 419-441; See also K. T. Ulrich and S. D. Eppinger, Product Design and Development. New York: McGraw-Hill, 1995.

15. Mikkola, J.H. & Gassmann, Oliver. (2003). Managing Modularity of Product Architectures. Engineering Management, IEEE Transactions on. 50. 204 - 218. 10.1109/TEM.2003.810826, p.204-205

16. Hölttä-Otto K, de Weck O. Degree of Modularity in Engineering Systems and Products with Technical and Business Constraints. Concurrent Engineering. 2007;15(2):113-126. doi:10.1177/1063293X07078931 Online: https://hal.archives-ouvertes.fr/hal-00571207/document

17. CXL Compute Express Link™. https://www.computeexpresslink.org/ At time of publication, the latest CXL revision 3.0, version 1.0 was dated August 1, 2022.

18. Leveson, N. G, Thomas, J.P, "STPA Handbook," March 2018

19. C. L. Liu and James W. Layland. 1973. Scheduling Algorithms for Multiprogramming in a Hard-Real-Time Environment. J. ACM 20, 1 (Jan. 1973), 46–61. Authors establish a set of assumptions for real-time scheduling, including using this synonym for partitioned tasks: "(A3) The tasks are independent…" https://doi.org/10.1145/321738.321743

20. Brown, M. Toward a taxonomy of communications security models. J Cryptogr Eng 3, 181–195 (2013). https://doi.org/10.1007/s13389-013-0058-2

21. FAA Technical Standard Order C153a, "INTEGRATED MODULAR AVIONICS (IMA) PLATFORM AND MODULES" 5/29/2019. Available at https://drs.faa.gov/browse/excelExternalWindow/DRSDOCID883243419720220303144951.0001

22. FAA Technical Standard Order C214, "FUNCTIONAL TSO EQUIPMENT USING A TSO-C153a-AUTHORIZED IMA PLATFORM(S) AND/ MODULE(S)" 7/29/2021. Available at

https://drs.faa.gov/browse/excelExternalWindow/72EB1A23D1AB3318862587210052652B.0001

Vertical Flight Society
2701 Prosperity Ave, Suite 210
Fairfax, VA 22031
USA

ISBN 978-1-7138-6505-6

CURRAN ASSOCIATES INC.
proceedings.com

9 781713 865056